THE SCIENCE OF LIFE
AND EVOLUTION

THE SCIENCE OF LIFE AND EVOLUTION

JOHN ADRIAN LEROY

Library of Congress Control Number: 2017903388
ISBN: Hardcover   978-1-5245-6130-7
                Softcover   978-1-5245-6129-1
                eBook   978-1-5245-6128-4

Print information available on the last page.

Rev. date: 03/03/2017

To order additional copies of this book, contact:
Xlibris
1-800-455-039
www.Xlibris.com.au
Orders@Xlibris.com.au
755271

CONTENTS

Nurture versus Nature (Rev 1 23rd Jun 2012) 1

Theory of Life and Evolution 4

Functioning of Cells..42

The Differentiation of Genes (Cells) 56

Life, Electricity and its Generation 63

06 Protein and its Role in Disease 70

EPIGENETICS, and the involvement of Chromatin
Protein in EVOLUTION...78

The fallacy of Stem Cells..83

EVOLUTIONARY DIVERGENCES as a RESULT of
CHANGED ENVRONMENTAL HABITAT EXPOSURE........88

LIFE, EVOLUTION and ENERGY PULSES...............................92

Function of the Auto-Immune System..106

LIFE, EVOLUTION and ENERGY PULSES............................. 112

Alzheimers, Dementia 126

Alcoholism 133

Autism, Its Cause 136

Penetration of Protein molecule by energy pulses.
(and cause of cancer) 143

Diabetes 151

The Cause of MS 161

Physical Dangers posed by Smart Meters,
Wind Turbines, Mobile Phones Etc...169

Protein, Smart Meters and Dope Taking by Athletes.....................177

NURTURE VERSUS NATURE
(Rev 1 23ʳᵈ Jun 2012)

A strange concept brought about by the prevailing ignorance of just what is life? How does it function? How does it begin? Where do our characteristics come from?

Answers to these questions (See Ref 1. for all the technical details) leads to the conclusion that nurture is the ongoing exposure of a specimen, during its lifetime, from the very first establishment of its DNA to its death, to the energy effects of its environment. (All effects of an environment on a species are by way of energy pulses causing a reaction from its DNA. These energy effects may be delivered by a secondary means, but they always access the DNA as energy pulses)

During the life of a specimen, if a persistent change of environment occurs the accompanying pulsating energy change causes excessive damaging vibrations of the DNA nucleotides pairs (Those initially tuned to the frequency of the changing energy pulses) resulting in a reaction and adaptation to this damaging effect with the installation of extended or further genes etc where etc is the junk DNA evolved to receive sensory inputs. These energy inputs associated with changed sensory effects with recognizable frequencies react with the DNA of the receiving cells (differentiated) devoted to the particular sense by causing recognizable distortions in the "junk" nucleotide pairs as they vibrate harmonically in response through the magnetic field present in the cells, sending on signals through descending number of cells where as a result, the final impression is coordinated in a receptor cell with all the relevant signals from the other associated sensor receiver cells. The resultant physical distortions are installed in the nucleotide pairs of the cell's DNA and when the installation of these effects requires a mental response, this is achieved through signals being generated through the neurotransmitter system.

On repetition of these environment energy effects the distortions becomes more established and a response becomes more readily available i.e. rapid memory response. Where the incoming environmental energy effects initiate a mental response and are persistent enough the effect is permanently installed in the cell's DNA due to straining of the involved nucleotide pairs by the stress effect of the repeated pulsing energy, as it reacts with them. The number of nucleotide pairs eventually being extended by a damage function resulting in replacement and addition of the involved nucleotide pairs with the capacity (tuned nucleotide pairs) to respond with necessary function and behavioural traits seemingly installed when activated by incoming information. This tuned arrangement of nucleotide pairs (strained) is eventually inherited by offspring making them responsive to incoming energy effects as were the parents and ancestors (Instinct).

The Genes associated with an environment change necessitating the physical adjustment of the species are adapted to provide an output of adjusted protein etc from the cell by installing in the individual's cells a further extension of the DNA (Regulatory Memory, (RM) and Genes) results in it coping with any of these damaging environmental changes (i.e. a healing effect required) and thus it, the species, with its DNA survives.

This process is briefly the simplified nurturing effect and it becomes more complicated as the species is evolved by its DNA reacting to the effects of the changed environment resulting in the production of proteins, hormones etc and mental outputs where applicable that account eventually for the functioning of differentiated cells and the characteristics of the species

This leads to the inescapable and logical truth that the DNA of a specimen is not necessarily constant over its lifetime but is always susceptible to evolutionary effects

These new Genes etc and RM (Junk DNA) associated with mental output or adaptations of existing genes etc established in a member or members of a breeding pair of the species, along with the pre-existing genes etc are passed on in accordance with Mendel's laws of inheritance and this results in the offspring reacting to an initiating input from this environment and in doing so demonstrating mental response characteristics and behavioural traits that are now seen as the nature of the specimen.

It is demonstrated then that nature is the result of the effects of nurture on the preceding generations of the specimen.

Conclusion

NATURE IS THE RESULT OF THE HISTORICAL NURTURING OF A SPECIMEN'S ANCESTORS AS THEY REACTED, VIA THEIR DNA, TO THE DEVELOPING ENVIRONMENT, RESULTING IN THE SURVIVAL OF THE SPECIMEN AND EVENTUALLY THE SPECIES.

Note

1. Nurture is not necessarily about parental influences, but as part of the specimens environment they may play their part in influencing (nurturing) behaviour, however the specimens nature and behaviour (Inherited) of any species results initially from the repeated nurturing effects during the ancestors' exposure to their environment.
2. Evolution is the result of the accumulation of genetic etc additions and changes in the species DNA by way of nurturing and inheritance as it reacts to the environmental energy effects of its exposure and is adapted to these effects of the environment, resulting in its survival. It is not due to natural selection or survival of the fittest.
3. Any environmental change must be slow and measured for the DNA to cope with the necessary adaptation.

Ref 1. "Theory of Life and Evolution" by John A. LeRoy (Unpublished)

THEORY OF LIFE AND EVOLUTION

◇◇

	SUBJECT	PAGE
1.	Alcoholism and Drugs, Inherited addiction.	Separate paper
2.	Behavioural traits, Inherited.	8.19.
3.	Brain and its cells	4.5.6
4.	Cancer, The causes	21.
5.	Carbon and Potassium, Radioactive (Role in life process).	2.3.
6.	Conscious memory	5.6.
7.	Chemically supported evolution.	2.10.
8.	Darwin's theories.	17.
9.	Dementia.	8.
10.	Diabetes. The causes.	Separate paper.
11.	Development of complex species (Eukaryotes).	10.
12.	DNA format.	2.
13.	DNA function.	2. 4.
14.	DNA imprinting (Installation of genes etc).	11.
15.	DNA Resonant body.	4.
16.	Embryonic stem cells (Description and function).	13.
17.	Energy, Environmental (Pulsing with fixed frequencies).	2. 4.
18.	Epigenetic effect (Detailed discussion).	Separate paper
19.	Epigenetic effect (Associated problems).	12.13.
20.	Eukaryote species	10.
21.	Evolution. (Definition).	2.3.19. 20.
22.	Evolution, system of cultivating internal cells of Eukaryote species	10. 14.

23. Fertilisation.   12.
24. Genes, Description and function.   3.
25. Genetic differentiation, how it functions.   3.
26. Genetically Modified food.   18.
27. Glial cells, their function.   5. 6.
28. Intelligent design.   17.
29. Jacob-Monod paper. E-Coli bacteria, how it   9. 10
     functions in reality.
30. "Junk" DNA.   4. 5. 6.
31. Life, description of and explanation   2, 8, 9.
32. Life, origin of.   2.
33. Methylation   16, 17.
34. Mitochondria DNA, role in evolution of   10.
     Eukaryote species.
35. Mammal species, reproduction of and evolution.   12. 13. 14.
36. Migration, cause of genetic incompatibility of   18.
     characteristics
37. Prokaryote species.   10.
38. PTSD.   7.
39. Regulatory memory, description and function   3, 4.
40. Sexual reproduction   12.13.
41. Species   11.
42. Telomeres and Telomerase enzyme. Origin and   10. 11.
     explanation of role.

HYPOTHESIS

Prior to the advent of life forms, the earth was consolidating with stable, recognizable elements with various properties and was a seething cauldron of these elements and energy.

Some four billion years ago the elements Carbon, Oxygen, Nitrogen, Hydrogen, Phosphorous and Potassium (radioactive) were available in various chemical formats. Traces of carbon (C14) included were (and are) radioactive and the conditions arose in a chemical rich swamp, which was conducive to forming complex molecules of RNA from nucleotides of the natural forming nucleotide acids adenine (A) guanine (G) thymine (T) cytosine (C) plus sugar and a phosphate compound.

From a pairing up of two compatible molecules of RNA, a molecule of DNA was formed. Initially the molecule would have consisted of very few pairs of nucleotides. The format of the DNA was (and is) a ladder like structure with the nucleotides being selectively bonded at their protruding extensions by hydrogen, forming ladder like rungs whilst the ladder like rails (uprights) are constituted of sugar into which one of the acids is integrated, forming a nucleotide. Bonding each nucleotide to the next in the rail is the phosphate.

When first formatted this organic chemical molecule had the three basic characteristics

(1) Being driven to be compatible with and therefore tending to achieve a neutral state within its environment. (As do all chemicals).
(2) Property of vibrating when pulsating energy from the environment accesses the DNA nucleotide pairs. (Vibration and resonance is the most common physical phenomena in the universe)
(3) The property of an extended life span when the arrangement of the nucleotide pairs is in harmony with the frequency of the applied environmental energy effects.

NOTE

The property of surviving for extended periods in pulsating energy effects and reacting to persistent environmental energy changes ensures the DNA's survival in its environment i.e. evolution. As it continually reacts and is changed to have an adjusted output compensating for the potentially damaging effects of the chaotically changing pulsating environmental energy, additional to the basic heat energy (always present) it makes the life phenomenon possible. It is the initial application of the pulsing heat energy effects that initially activates the DNA when it is formed, and as a result of the DNA being exposed to additional environmental effects genes are formed in sequence as they are established resulting in them being available for activation when the relevant energy pulse is present, bringing into play the relevant cells and differentiation.

At first the probability is the initially formed DNA would have occurred from the final uniting of the RNA chemical molecules in a soupy mixture of water and chemicals including sugar. Here it was initially exposed to the environmental energy (heat) and was reactive as a body to the resonating influences of the pulsating energy. This initial response to the heat was the effect (existing today) from the DNA molecule as it formed, causing it and therefore the resultant evolved cell to be susceptible to the resonating effects of the energy of its environment. The vibrant effects on the nucleotides continue throughout the life span of the resulting cell, varying as the strength of the incoming energy changes. Exposure of the rung like nucleotide pairs of the DNA to the basic pulsing (heat) energy of the environment and the subsidiary energy effects of the environment causes stress raisers and weakening of the hydrogen joints of the rungs. Depending on the number of energy cycles they have been exposed to fatigue occurs and the DNA is split apart at the weak point, the hydrogen joint. An enzyme assisted chemical response rebuilds the two halves of the DNA chromosome resulting in a doubling of cell numbers.

As conditions on earth changed i.e. light radiations, colour changes etc additional energy inputs began to complement the basic energy and to access the DNA molecules. These normally encountered environmental energies were compatible with the dimensions of the nucleotide pairs of the DNA. (In more advanced species the energy pulses may be provided by a secondary means as the sustenance induced chemicals are broken down in the cells, releasing energy pulses commensurate with the environment the sustenance was produced in). Certain of the nucleotide pairs of the DNA however were now not harmonically responsive to this new energy frequency and when the additional energy effect was persistent they sustained damage and were destroyed. When this occurred a healing process to the DNA resulted as enzymes and proteins formed suitable for the production of the nucleotide acids and an enzyme aided chemical response "damage controlling" the effects of this energy, brought about the replacement and addition of nucleotides pairs that were once again harmonically tuned (Regulatory Memory, RM) to the frequency of the energy pulses and on the application of further bursts of this environmental energy effect (Applied chaotically) they responded by emitting an energy signal reflective of this fluctuating energy effect. This response of the RM identified further damage, but not destruction, that were confined to adjacent nucleotide pairs that were strained beyond their elastic limit (The Gene). This damage now reflected the damaging capabilities of this energy and identified the chemical reaction necessary to control further damage, leading to the survival of the DNA (and development of its species) in this changed environment. (As below)

The RM reaction is due to the relevant nucleotide pairs vibrating through the magnetic energy field provided by the radioactive C14 and Potassium (K40) included in most species cells. (Or rarely in some other circumstance an alternate magnetic field provider) generating an alternating electric current

NOTE

Life does not and cannot survive without this magnetic field phenomenon. Potassium, an abundant element in the earth's makeup has an isotope, K40 that normally provides this energy field, however in the rare instance of the potassium not being present and life existing there is always another source of the magnetic field present in the cells of a life form.

The fluctuating force field (switching/signaling) surrounding the pulsating electric current generated in the series of RM nucleotide pairs identifies the gene(s) resulting in the initiating of a reaction (mRNA) from the gene(s) etc that is a guide for the cell to produce an appropriate protein(s) etc reaction, resulting in its survival in this environment. This characteristic of the DNA results in its survival and it never varies in reacting continually maintaining this state in any normal changing environment (A non permanent chaotic effect) it is being exposed to (Including prevailing environmental affects). Eventually if persistent changes to the environment energy effects take place destructive damage to now non-reactive DNA nucleotide pairs occurs and the process to achieve survival is repeated. This is evolution as the various proteins etc produced are utilized to develop the species, ensuring the survival of the DNA.

The electric current generation is restricted over a very small length of the DNA (Ref 2.) thus ensuring the generated current and its resulting energy field is restricted and not dissipated as it identifies the relevant gene thereby gaining a response. This fluctuating energy field then highlights the damaged nucleotide pairs that have changed electrical properties, (the gene) and guides the formation of an RNA messenger that guides a compensatory enzyme action in the production of the required protein etc from a ribosome, based on the potential damage effect and this supports the DNA's survival. The fixed frequency resonating effect of the incoming environmental energy resulted initially in damage to the nucleotide pairs, causing the RM to form and creating the gene. This identifies the variation of this energy from the original resulting in the establishment of a formula of proteins, hormones, enzymes etc required to sustain the DNA and maintain the cell against damage from this specific environment affect. Over the course of its evolution the species DNA has developed specific cells in response to these environmental energy effects with the cells outer envelope (membrane) having receptors evolved to cope only with the energy sources that are relevant to its evolution and function. When the energy pulses within the cell are released or applied they identify (or activate) the genes etc (possibly an enzyme action accelerating the formation of the mRNA) they are tuned to. This is "Differentiation" of the genes. The ongoing reactions from the relevant RM's to the environmental type energies designates, via the genes, by mRNA action the required enzyme, proteins etc, and due to the fluctuations of the environmental energy input selectively activating the identified mRNA, the control, disposition and quantity of the output from the cells necessary to enable the establishment, maintenance and survival of the DNA and hence the evolutionary generation and support of a species.

The products from multiple genes, evolved in response to a multitude of incoming environmental affects can be seen as virtually scab tissue protecting the DNA i.e. it is the species, and it is this process that ensures development of varying species and hence their DNA's survival in a variety of environments.

After the development of the eukaryotic species, due to the many environments and environmental changes undergone the resultant effect on the DNA from the variety of damage inflicted and the consequent healing reactions has developed many species, some evolving brain cells in parallel with sensory organs resulting in the mental ability and control necessary for mobility required to access sustenance. This characteristic occurred due to the evolution of the DNA, establishing the process of the species being mobile and thus surviving in the environment. The evolution of this characteristic is due to the original property of the DNA being a molecular body where an arrangement of designated nucleotide pairs (Regulatory Memory (RM)) associated with a gene is tuned to resonate harmonically to specific incoming environmental energy pulses (Ref 1.) and thereby "switch" a response from the gene. When, due to the necessity to access sustenance, the required mobility meant the upgrading of the DNA's characteristics of apparent logical control, thinking and aggressiveness to survive along with its capacity to evolve the physical characteristics necessary, the RM nucleotide pairs and genes evolved to respond to this more complex environmental energy input and thus the output gradually became more sophisticated, producing the effect of initiating a controlling organ, the brain and fins, arms, legs etc supported by the parallel evolution of the sensory organs.

The conscious brain cells were established, integrating with the cell communication system, (Neuronal cells) in response to the incoming reflected energy effects indicating the presence of sustenance at a distance, establishing and activating (Evolving) RM nucleotide pairs capable of initiating a reaction from the DNA and hence its cell. Cells were evolved capable of delivering energy control signals to evolving muscles etc via nerve fibres. They also evolved to conduct initiating control signals between relevant cells with an intermittent protein connection depending on the frequency of the incoming energy relevant to this environment event requiring a conscious reaction. An output of facilitating protein to fill the "synaptic" gap between the evolving cells (Glia) and an increased energy signaling (thinking and conscious memory) ability to control movement by increased cell numbers was a parallel development as the Glia cells interfaced with the neuronal cells.

As the species ability to move led to slight changes in its environment further damage to the nucleotide pairs involved with a change of energy input resulted in a survival and healing effect extension of the number of responsive nucleotide pairs representing an increased array of incoming information to be installed (memorized) and responded to. As the environmental energy pulse information is received via the evolved sensory organ (Ref 6) they are delivered to associated receiver cells resulting in an energy signal of a frequency allowing recognition and sent on to the cell's DNA of the appropriate brain sub organs. If the information is of a repeated nature and has previously been installed in nucleotide pairs as a physical recognizable distortion (memory affect) the format of the original energy pulses and these repeated pulses are the same, they eventually access and act on the relevant cells DNA nucleotide pairs (The DNA and hence the cells are very specific in the function they perform (Ref 3.)) by the energy pulses accessing and further distorting the possibly already affected (Memory installed) nucleotide pairs. The reaction to this is a conscious effect in recognizing the event and initiating a survival reaction. These nucleotide pairs of the DNA "The Junk" along with sensory inputs have developed in the DNA in an arrangement making them harmonically reactive to the incoming energy pulses representing an event. The more persistently particular information is incoming to these specifically developed nucleotide pairs the greater the degree of strain distortion becomes on them and this represents an event or object etc. As the more often this exposure to an event or to an object etc occurs the degree of the relevant strain increases in the nucleotide pairs and the longer the recovery time and therefore longer lasting conscious memory. (These cognitive permanent memories eventually become installed in the neuronal cells making for extremely rapid response in the event of danger etc). This system results in the brain not becoming overloaded with information. As the relevant nucleotide pairs are vibrated during these events they create signals (energy pulses) from the cells that are relayed to a mental response centre coordinating mental and physical reactions. Persistent variations, and extension of the incoming energy influences via the senses causes permanent strain of the "junk" nucleotide pairs inducing the DNA to guide the production of enzymes that then assist in the function of initiating the development and extension to the DNA of additional "Junk" nucleotide pairs thus sequentially extending the DNA chromosomes of the relevant glial cells and therefore increasing the brain capacity. Associated organs and system capacity evolve in parallel with this increased capacity. If occurring prior to the conception of an offspring and both parents are exposed then this increased capacity may be passed on. (Evolution)

This system results in the brain having great memory and reactive capacity, established in degree depending on the species exposure to environmental energy influences and as the DNA of a cell is very specifically oriented to a task the brain cells have evolved, coping with the required numbers of the different types of incoming energy pulses. As a result of the "pressure" of the incoming energy pulses, applying to individual cells the cells duplicate and the organ keeps growing until the number of new cells of the organ equalize with the demise rate of cells as they disintegrate due to the pressure of their work load. In parallel with the development of the brain organs, sensory organs evolved to cope with the incoming environmental information and to process it in a controlled manner via the relevant limbs, organs etc. The brain has evolved with two basic capacities.

1. To control the ability to balance the body in action and in the human this occupies approximately half of the brain cells, with connections to the conscious and mental outputs of the brain. This characteristic in mammals is associated with the ear organ.
2. To provide the conscious and mental outputs initially evolved to control mobility via limbs etc but latter to facilitate thinking. (A matter of degree)

The reaction in achieving the second capacity can be due to one of three methods of receiving the information in the form of energy pulses relating to the environment.

First, via all sensory sub organs, consisting of relevant action initiating cells receiving the information then sending the cells responses representing the environmental energy input to the brain sub organs to collate the incoming information resulting in memorizing

Secondly a conscious mental output from the brain sub organs initiated from an input, accessing the neuronal cell system, originally evolved to cope with communication between cells, but later developed by an evolutionary process to cope with the species mobility and thus survival.

Thirdly utilizing the information and mental capacity (Evolved from the initial "awareness" ability of the DNA) to initiate the mental responses. (Thinking, Logic.)

The Glial system has evolved with sub regions of the brain devoted to the various sensory systems and these consist of millions of particular sensory cells, along with others (Normally descriptively lumped together as Glial cells), specialized for maintenance tasks. A species cells have been evolved selectively, due to the various ranges of environmental energy effects, with the resulting effect of survival. This includes sensory cells evolved with the capacity to react to the environmental energy related to the particular sense and to communicate this to the cells of the sub organs evolved to display the characteristics of consciousness and mental capacity (These characteristics involve memory, as every other characteristic in any species does and the capacities for these memories are initially installed in the DNA, by the replacement of strained and destructively damaged non-harmonic nucleotide pairs' with pairs that are now compatible with a changed or newly encountered energy pulse of a different frequency. This damage initiates an enzyme assisted renewal and additional extension of nucleotide pairs resulting in the increased capacity of the DNA to recognize and respond to the stressful energy pulsing process, where mobility is indicated for survival, at the continued repetition of this energy effect).

The sub organ of the brain receives all these congruent energy influences as they affect the specimen and are sent together to these organs and thus they represent the environment as it influences the individual. The cells of the eye sensory organ, for instance in humans, receive energy pulses constituted of a mix of the primary colours ranging across some two million shades of colour. The output from these cells, each of which has the "Junk" portion of the DNA developed to receive incoming colour frequencies are sent on by a radiated energy process in the form of pulsating energy of a fixed frequency for a colour, where receptor cells in a screen like arrangement recognizes, via the "junk" DNA nucleotide pairs this spread of colour as it accesses the harmonically responsive nucleotide pairs. This particular ability of the human has evolved as a response to the range of colours, assisting in its survival. The receptor cells react with an energy output representing the colours with an integrated like response virtually simulating the "Pixel" output of a TV screen. This integrated with the outputs from the other senses enables an awareness to eventuate and this is consciousness

When born a child does not have the colour effects installed in the brain cells DNA but have evolved "junk" nucleotide pairs capable of accepting the impulses related to the colour. Gradual repeated exposure in the first few years of life installs these colours to all of the relevant brain sub organ receptor cells. The incoming effects from all the senses are installed in a similar manner. Repeated exposures to environmental events, objects etc, involving the installed information in all of the cells, over these first few years of the child's life results in rapid concerted energy outputs from the cells when further exposed to energy pulses due to continual fluctuating environmental events, that the child consciously recognizes, due to the awareness of the DNA, as it strives for survival.

When received by the sensory organ cells the strength of the energy pulses of any particular shade (Colour) or other sense is sent on to the relevant cells where it is received in this condition resulting in an adjusted output that gives the perception of distance and aids in the 3D effect resulting from eye disposition.

Note

1. Every image registered is in fact responses to energy impulses representing various colours and strength of signals and depending on the strength of these signals the variation in response signals a 3D spatial effect to be created. All of the sensory organs inputs are received in a similar manner adding to the spatial effect and when united represent a physical depiction of the species environment.
2. As a further response to this acquired characteristic of the species registering its environment to identify sustenance, it was necessary for it to be mobile and the property of awareness of the DNA that drove survival of its cells in its environment evolved into a mental capacity, initially to access limbs etc and thereby controls movement.

Mental thinking and reasoning output appears to involve the ability of the DNA to respond to an incoming initiating sensory environmental influence. Then due to the characteristic of DNA of reacting to any input energy, a search sequence is set in action through related brain cells as it sifts the contents, eventually coordinating the results. In humans this is a throw back to the characteristic of logic controlling movement and therefore survival. Much of the mental response capacity is based on memory of past events that are, as with any other memories installed by the resulting energy inputs emanating from the event. This then indicates that as the DNA always responds with the same basic process i.e. environmental energy input triggers a response from the

DNA by utilizing energy with a resultant output aiding in the survival of the species, these energy outputs from the myriad of cells involved in the sensory receptor cells result in an overall sense of the environment which is consciousness that is then utilized by the mental capacity to reach decisions as necessary.

The Astrocyte Glia cells communicate with other cells, including Neuron cells by radiated energy pulses, initiated by the incoming energy pulses from the environment via the specific sensory organ, in an electro chemical process. This system applies to all of the sensory organs. There is a specialized brain sub organ with cells that have the function of receiving this information, processing it and activating physical responses, via the neuronal cells and nerve fibres. This system involving the neuronal cells deals with both the reactions necessary to cope with the incoming environmental effects and providing outgoing signals and returns from the limbs and organs required to control physical responses necessary for survival. These return signals are fed back into the Glia cell system, invoking awareness of the reactions then activating follow up actions to cope with the environment dictates.

The rapid reaction characteristic in time of danger etc is due to the involved neuronal cells, being available for extremely rapid access to receive and relay signals to control physical movement as necessary. This occurs as the relevant cell is subjected to persistent and sometimes stressful signals as the incoming frequency of energy signals are recognized and passed through restricting previously established pathways, eventually being installed in the cell and retaining the effects of the environmental energy pulses (Distortions of nucleotide pairs), due to an image, that culminates in a long-term memory being installed. The result of this is the processing organ of the brain does not have to search the memory banks and thus rapid responses required for survival become available. Logically, the forerunner of the neuronal system, by necessity, was in existence prior to the establishment of mobile species, enabling some awareness between cells of immobile multi cell species.

Various glial cells have been evolved to provide the capacity to gather (memorize) and maintain the process of receiving environmental inputs related to mobility and the neuronal system has been adapted and utilized to evolve cells and nerve fibres to distribute action signals. There are also cells in the brain employed in reprocessing the protein that is initially injected intermittently on demand into utilized cell synapses, and then eventually discarded, if the involved incoming information is not repeated.

The Synapse is a ball and socket like connection between brain cells where the connection is a gap that is activated by an injection of protein formulated to convey specific information signals when required. When the glial receptor cells associated with the various receptor sensors are energy activated a response of a protein capable of transferring these resultant signals across the gap between the downstream cells and the specific neuronal permanent memory cell is injected. The communicating cell synapses, relevant to an incoming event are sensitized by the injected protein being formulated in the injecting cell by the DNA's response to specific frequencies relevant to the image of the object being installed. If the image of the object being installed is repeated frequently then the sensory cells output of associated radiated energy impulses are recognized by the previously associated glial cells activating a repeat in the receiving neuronal cells where the image is directed via the previously established protein capable of transferring the information. This eventually leads to the build up of protein of the same formulation and establishment of permanent pathways of access to the neuron cell recording the event where all relevant facets are installed. The more often an event leading to an energy pulse transfer in the synapse is repeated the more consolidated the protein becomes in the synaptic gap. (Ref 5), and available, where the relevant receiving nucleotide pairs are eventually distorted requiring long term recovery (Long term memory) or strained beyond their elastic limit representing this memory, enabling instant recall, and action, when prompted by a recognizable energy pulse input.

This system allows for various intermittent environmental effects to be registered as required without overwhelming the brain, with neuronal cells remaining available and generating a response but recovering and being available for responses when required, whilst committed cells are available for oft-repeated images. (Danger threatening events accelerate the permanent installation of the memory, possibly with such chemicals as Serotonin, Dopamine or Adrenaline adjusting the response. These proteins are produced as a genetic response and aid in the control of reactions to specific events.)

As described above the advent of an environment requiring mobility, cells capable of controlling this movement evolved with the characteristics of apparent thinking and logic based on the property of awareness responses of the DNA leading to survival in its environment. To apply the output of these cells established in response to incoming environmental effects i.e. energy pulses reflecting sustenance out of reach, they evolved as a component of a relay system, operating via the neuronal (communication) system to establish physical control via evolved (in parallel) fins, feathers limbs etc. (Glial cells were previously thought of as extraneous cells, this is not a viable proposition as each and every cell is evolved due to the incoming environmental energy with resultant effects being the number of cells available, as the production of new cells stabilizes with the cells being destroyed, dictates the overall reaction of the mature organ etc).

Evolved in parallel with the sensory organs, due to the development of changed environmental energy input stresses necessitating a mental output, increased numbers of these cells, along with maintenance cells in the brain (known and grouped together as the Glial cells) have been evolved forming dedicated sub organs of the brain, including sensory receptor organs. Purpose functioning cells of the Glia, the Astrocyte were evolved to interpret the various types of inputs from the senses, resulting in outputs from these cells, that when coordinated represented the physical environment, required to assist in the establishment of physical control of the specimen.

Depending on the range of environmental effects the mobile species was exposed to, increasing numbers of cells were evolved and with the increased ability of mankind to create situations, adding to and adapting their environments, it resulted in mankind's brainpower (cells) being evolved in huge numbers of Glia cells to cope with these hugely increased numbers of environmental effects, far outstripping the numbers present in the next most intelligent species by a factor of nine.

Incoming information of a frequency sequence, often repeated results in a neuronal cell becoming dedicated to that subject (Ref 3.) with dedicated nucleotide pairs of a cells DNA gradually becoming more strained with a longer recovery period and finally becoming a permanent memory, when the straining of the DNA nucleotide pairs is established beyond their elastic limit. (The agility of any species relates to the number of glial and neuronal cells and hence its mental capacity).

This is the activating process that has evolved leading to survival of the species in an environment requiring mobility, however with an increase in the number of cells and the action potential required, secondary systems in control of the primary system have been evolved in mobile species providing the necessary energy (ATP) and electrical conductivity (Calcium) to support the process.

The apparent breakdown of the receptor cells, with incoming memory and information function is induced by the inability of the dedicated ageing protein recycling cells to destroy the excess protein. This leads to the early cells in the memory train becoming dysfunctional and dying resulting, by degree, to the inability of the mentally processing and short term memory cells to receive and activate new incoming information and short term memories, whereas previous and permanent long term memories are still available from the neuronal cells i.e. Dementia. It also leads to the death of the protein recycling cells as they become choked up. The replacement of support cells in the brain is initiated by the cell duplication process i.e. the aged cells divide and replace "for purpose" cells. The replacement of these cells therefore cannot be achieved as the potential duplicating cells are demised.

When repeated events causing permanent conscious memories to occur i.e. suckling, before reproduction of an offspring, the effects of the influences causing them, are also installed in the reproductive DNA of the female cells (eggs) and the male reproductive organs producing sperm, are passed on in part of the process of evolution. (These memories are known as "Instinct" but they are not, they are inherited memories controlling behaviour, possibly passed down as non-coding distortions of the DNA nucleotide pairs in the Junk portion of the DNA).

Evolved brain cells have the property of inherited permanent "Instinct" memories and the evolved capacity to retain a range of memories with the ability to utilize these to achieve a thinking output as a response to environmental energy inputs. Designated cells evolved specifically for the purpose are activated to produce controlling chemicals such as serotonin, dopamine and adrenaline etc that either muffle or exaggerate the effects of the incoming signals as necessary to achieve survival responses.

Brain cells can process the recall of conscious memories and thinking due to the evolved property of awareness in the RM of the DNA (A coordinated, concentrated effect dependent on the billions of cells devoted to this process) and in the event of a horrifying or depressing event recalling them repeatedly and this can permanently install this memory in the relevant DNA and this accounts for Post Traumatic Stress Disorder (PTSD) and if the effects are severe enough, resulting in permanent installation descendants can inherit the depressive traits as clinical depression.

The number of "Junk" nucleotide pairs and the number of cells coping with the mental response characteristic is relevant to the evolved agility of the species (The human is the most agile species) with the number of "Glia," cells evolved coordinating with the already existing communication system, "The Neuronal" in multi cell species as complementing characteristics necessary for mobility evolved, i.e. as well as the senses, fins etc and eventually limbs.

The greater the agility of the species due to the environmental pressures the better the evolved conscious memory is. For the human species this memory characteristic led to the extension of its environment and further learning and therefore extension of the memory capacity until not only are advantages to the human species available, but there are also some significant disadvantages. This process is an ongoing part of evolution.

The more complex an environment becomes, the more complex the incoming environmental effects (energy pulses received) become and this drives an escalating rate of evolution of the species' related characteristics.

At the beginning of the life process the DNA molecules produced as above were activated and whipped around due to the vibrating effect of the existing environmental heat energy pulses. Eventually the whipping response to the energy caused the ends of the DNA to come in contact and join up forming a circular band of DNA, and the basis for life and the prokaryote species was established. (As the formation of the Telomere (Protein) a scab capping of an ending of a DNA strand (chromosome) in eukaryotic species is a time consuming process this situation did not occur initially and allowed the ends of the DNA strand to join, forming the circular DNA structure of the prokaryotic species).

The RM and gene responses are as described previously, however as they were initiated as prokaryote species and were virtually dependent on direct energy exposure little progress in the way of species evolution eventuated in the first three billion years. The DNA eventually evolved a cell producing the protein etc required for the survival of the DNA in this environment by continuing to chemically reduce naturally available sugar, thereby releasing the energy requirement to produce chemicals from available elements for repair and replacement of the cell and chemicals, protein etc for a protective framework (The species) resulting in survival of the DNA with an energy supply that was additional to the direct available environmental energy.

Gradually from the beginning and the passing of time, persistent changes to the initial environment occurred with additions to the environmental energy, resulting in, if the addition was a closely allied variation of an energy effect a modification to the existing gene and RM and if a significant change an additional gene and RM with an accompanying expansion of the DNA molecule developed as described above. (These changes resulted in the production of protein closely allied to the DNA and its exposure to a particular wave energy environmental effect. This protein then has the characteristic of ensuring the DNA survives in this environment effect (A "scab" formation)) Gradually as the DNA developed the capacity evolved to provide protection in the form of the cell membrane and more efficient methods of ensuring survival i.e. organelles etc involved in the production of hormones, enzymes and proteins etc relevant to its survival. An additional capacity also evolved to establish limited sugar reducing molecules of ATP involving the cell membrane, providing an additional source of energy.

The process of evolution for prokaryote species involves a direct reproduction of the cells i.e. daughter cells followed by daughter cells etc and hence any changes in the DNA incurred over the specimens life time is directly handed down making the pace of evolution due to outside influences much more rapid than for eukaryote species who are evolved in accordance with the principles laid down in Mendel's Laws of Inheritance, an indirect system. The prokaryote species however, because of their initial state of fragility and lack of a continuous supply of energy are normally exposed to restricted environments and this has slowed down their development.

An illustration of DNA and hence life forms to be evolved by different environments is the E.Coli bacterium, a prokaryote that has been evolved to survive in the gut environment of mammals. Initially when the ancestors of prokaryote species came into being no eukaryote species (mammals etc) existed and this provides proof life forms can be evolved by DNA being subjected to an energy input from a changed environment.

NOTE

The paper by Jacob–Monod discussing the functioning of the advanced E.Coli bacteria in the gut of mammals when surviving on lactose as the offspring is suckling, obviously lacks an initiating process to kick start it and this is provided as follows. The evolved repressor protein molecule in the absence of the lactose (A conduit for environmental energy) blocks the nucleotides of the operator gene from vibrating and therefore rejecting a positive signaling start to the genes, where no lactose exists. In the presence of the lactose this protein initiates a chemical response between them releasing energy pulses that activate the now freed operator gene nucleotides that then sets in motion the process to signal the structural genes that enzymes are required from the ribosomes to initiate the digestion of the available lactose. As the process of the lactose/enzyme reaction releases a steady stream of pulsating energy it keeps the process in action as it activates a continuing signal response from the operator gene (RM?). When the lactose supply diminishes the energy pulses diminish and the regulator gene guides the reproduction of the repressor protein molecule that locks down the operator gene.

The above illustrates the beginning of the life process with the development of prokaryote species and the following describes the second phase, the Eukaryotic

One billion years ago? a prokaryote species (The mitochondria) was evolved with the capacity to initiate the production of many molecules of ATP via its DNA resulting in the capacity to continually produce pulses of energy by chemically reducing increased amounts of naturally occurring sugar, in a controlled process without the species being directly exposed to environmental energy attached itself to other prokaryote species in the manner of a virus.

This "infection" occurred in an indiscriminant manner, with various numbers of the "virus" like species attaching to other prokaryote cells. These bi-species then began to evolve more rapidly (duplicate cells), due to being subjected to a continuous supply of energy, as they became more sensitive to changed environmental conditions with a more rapid increase in the DNA molecule length (Increasing due to the additional variety of the now continuous environmental energy affects it was being subjected to day and night, resulting in it extending with genes etc increasing the amount of information able to be stored (recorded) and responded to). This development led to increased sensitivity to change and the evolution into species with cells no longer exposed to direct environmental energy pulses.

As evolution of these more complex prokaryote species was taking off some were further infected by other prokaryote species without the energy function of the mitochondria, with its DNA penetrating into and through the original cell membrane. This resulted in a doubling of the DNA within the cell membrane with the two DNA components having slightly different features, however as they underwent further changes to the environment they were being evolved similarly as they were being subjected to the same environmental energy, resulting in the possibility of one of two species with mostly similar features and characteristics being produced with a statistical chance of 50/50 for each of the species from a single cell. Due to the continuing exposure of the species to additional energy and the establishment of survival characteristics the species evolved into multi cell units with the doubling of the DNA within the cells. The reproduction process present in the original prokaryote species ensured that the DNA was upgraded as the environment changed and this process still existed. To maintain this survival characteristic of renewal the process reverted to the original system by producing a cell with a single set of DNA complete with the mitochondria DNA, that was based on the prime prokaryote species and then an additional cell based on the infecting prokaryote species that was then free to infect any similar prime cell thus repeating the process.

As evolution proceeded a specimen exposed to the increasing stress in the increasingly tightly wound coils (2) of the DNA molecules in the cell's confines led to strain of the circular DNA molecule and the eventual rupturing of the DNA molecules into various numbers of matching strands, the forerunners of chromosome pairs.

Survival of the DNA is the prime characteristic and as the format of the cell was now unwieldy with the chromosomes and ribosomes etc now clogging the original cell membranes evolutionary pressure caused changes to the cell and a more efficient process eventuated. The original membrane evolved to be the Nuclei membrane of the eukaryote species and the internal functioning machinery (Organelles) evolved to be housed in the cytoplasm along with the energy supplying "Bacteria", the Mitochondria.

As the species continued to evolve and multi cells were developing as they were exposed to natural sugar, the incoming heat energy released from the sugar (The secondary means) continued to activate the DNA chromosome molecules with vigorous vibration resulting, causing the destruction of the end nucleotides from which an enzyme formed capable of re-generating the end nucleotides and initiating the production of a protein (The chromosome ends did not reunite as they were too tightly packed). This protein coated the ends of the molecules forming a protective scab, stabilizing the chromosomes. When the reproduction of the cell occurred the chromosomes of the daughter cells were lengthened as a response to the rebuilding enzyme action and they were then once again reduced as the process of cell activation and reproduction was repeated requiring production of further enzymes.

The protein "scab" capping are known as the Telomeres" and the enzyme as "Telomerase".

As the cells are diversifying and the DNA being reformatted this initial process of the application of basic energy is a damaging effect and not only causes the Telomeres to form but also a coating of protein, the "Epigenetic" effect (The outer coating of the protein pieces), is formed around each chromosome, muffling the agitating effects of the incoming environmental energy pulses leading to stabilization of the chromosomes. The basic heat energy effect continues for the life of the cell keeping it in an active (vibrant) state, whereas specific genetic and mental responses are called up when a variation to the environmental energy affecting a reaction is present.

A logical examination of the record of DNA shows that the operating principles never change but it adapts and develops more sophisticated versions of these principles. Having conceived that life began with the application of environmental heat to the initially short length of DNA causing a reaction when it was activated by the pulsing energy it is not inconceivable that the virgin DNA is still activated and continues to be, in all cases, by the application of environmental heat energy, by direct or indirect means and then its survival is dependant on its adaptation to the energy variations of its changed and changing environment. The importance of maintaining the Telomeres endings then is that when basic energy is directly applied to the DNA and is activating it

the endings conserve the chromosomes, whilst the relevant genes etc
support it by supplying the necessary adaptation for survival in the
changed and changing fluctuating environmental energy fields. The
telomere scabs then prevent further damage to the chromosome ends
and enable the process to survive

When cells divide it is due to the vibrating energy applied over a
significant period resulting in damage due to the number of energy
cycles the hydrogen bonds of the nucleotide pairs of the DNA have
sustained (fatigued) causing the DNA to split in halves with an enzyme
aided chemical response producing complimentary sides of RNA thus
doubling the DNA strands and multiplying the cell numbers.

The DNA having reproduced as a result of the applied energy
causing splitting of the original chromosome strand into RNA like
strands and complimented with newly produced matching strands of
the chemical molecule of RNA does not and cannot have the genes and
information contained in the originating DNA installed chemically,
but the relatively severe vibration of the DNA caused by the incoming
heat energy that initially activates the DNA of all cells in some form or
another and therefore vibrates all the nucleotide pairs of the DNA of the
daughter cells installs them. On the half strand from the original DNA
the nucleotides depicting the survival information (genes) are slack and
fatigued and therefore offer no resistance to this pulsating energy and the
matching nucleotides when paired up with them, without the normal
resistance (stiffness) quickly become fatigued when exposed to the
activating energy and the new DNA then contains all the information
of the pre-existing genes.

With the increasing crowding within the nuclei of the cell due to the
epigenetic effect and expansion of the chromosomes with the additional
information resulting from the increased incoming environmental
energy affects (As species evolve, their actions add to the complexity of
their environment and this increases the environmental affects on their
evolution) protein structures (Histones) have been evolved, exposing
the relevant genes etc of the chromosomes, to the relevant incoming
environmental energy. The epigenetic coating affect however, because
of its nature of production has a propensity to be uncontrolled and
can, due to the vibration and crowding of the chromosomes and the
epigenetic volume within the nucleus membrane may sometimes cause
malfunctioning between the dominant genes etc and the incoming

relevant energy pulses leading to dysfunctional performances such as schizophrenia (Along with the phenomena known as "Chromosomal crossing over" may cause confusion of traits, characteristics, as the relevant DNA nucleotide pairs have been evolved in different environments) The "Epigenetic" effect does not and cannot have a reactive capacity and any problems occurring due to it is the result of faulty disposition. The familial occurrence of these problems may be due to an inherited gene mutation tending to cause a buildup of epigenetic protein in a particular area

As the transition process from prokaryote to eukaryote species was evolving due to the invasion of one prokaryote species (A simple species, uncommitted to any characteristics except that of survival by invading another and a few other survival supporting characteristics) by the other resulted in the possible production of either one of a specimen of a complementary species. The addition of one prokaryote species to another's cell resulted in the situation where a doubling of the DNA content within the cell membrane led to the reproduction (daughter cell) process being activated (fertilisation). This process has evolved into the sexual form of reproduction with the adaptation of the pairing of chromosomes ensuring more complex species with the ability to survive in more complex environments.

On examination it can be observed that the sexual process mirrors that of a prokaryote cell invading another and pressurizing the production of a species with many daughter cell reproductions, utilizing randomly selected chromosomes from the available pairs of each parent to establish mixed characteristics with the pertinent characteristics of either female or male being normally dominant.

The process is not an extremely precise one and at times problems can occur resulting in what is known as genetic dysfunctions. The process results in compatible species (female and male) functioning generally as complex prokaryote species, carrying within their chromosomes and genes the possibility of statistically reproducing an enormous number of variants of the individual species. dependent on the environmental exposure history of their predecessors.

The details of the reproduction process for mammals is as follows:

For the female, as it was when the eukaryote life process began the initial cells (eggs) produced by the maternal partner are in the prokaryote form of species, and are complete with the mitochondria DNA that originally invaded them.

NOTE

The gametes are produced during the fetus stage of a female mammal embryo and these are carried throughout her life, being released selectively to be available at the sexually active stage. The gametes have only singular X chromosomes selected randomly from each pair of chromosomes making up the maternal specimens original DNA chromosomes. These cells (gametes) are stored in the ovum organ of the female until made available for the reproduction process. The gamete' yolks are supplied with activating chemicals and proteins etc. required for the building of and activating of the embryo cells on application of appropriate energy pulses. These chemicals have evolved as a result of the environmental evolutionary processes the female's ancestors have been exposed to and are installed from the female as the gametes form. These chemicals drive the differentiation process of the eukaryotic species cells and if the female, prior to conception occurring is subjected, similarly to that occurring in any somatic cell, slight variations of persistent environmental energy pulses the resultant bio-chemicals have been adapted resulting in an adjusted output from the developing offspring's cells leading to evolution and survival in the changing environment. If these chemicals, on chemical reduction release energy pulses with a frequency compatible with the RM and genes etc. of the developing cell's DNA chromosomes nil effects occur.

Further chemicals that are involved but evolved at a later stage in the species ancestral history are eventually supplied via the blood supply of the placenta. These chemicals are now supplied as they are to a normal adult's cell and are involved in the activation and development of every type of differentiated cell throughout the species. However if one of these chemical has been modified due to a persistent environmental change the DNA, genes etc. of the relevant developing differentiated cells are updated, as the revised energy pulse frequencies change, preparing for the specimens relevant cells to be compatible with the changed environment i.e. Evolution. Any number of changes can be made in this manner and passed down to the offspring. Direct energy effects, via sensors i.e. eyes ears etc., involved with mental inputs, if persistently repeated, have their environmental evolutionary effects installed into the DNA in an ongoing basis.

A similar process as above also applies to the cells of the male reproductive organ giving it the capacity to install in the sperm (The sperm of the male human may carry an X (Female) or Y (Male) chromosome with any ongoing evolutionary advances in the DNA genes etc the male has been subjected to during its preconception lifetime as well as all existing genes in the particular chromosomes that have been allocated to the sperm.

On fertilisation, once the sperm is induced to the gametes the DNA chromosome pairs are in a state ready to be activated on the application of heat energy pulses. This heat not only initiates the multiplication of the cell to the embryo state it also induces the cell's randomly allocated chemicals to be broken down releasing energy in a series of pulses. This pulse effect supplies the additional activation (switching) to the genes etc as the incoming heat energy is responsible for the vibrant characteristic of the DNA and with the activated genes guidance the output of the cell and consequently the developing embryo's life is underway. As the developing specimen progresses the it's cells produce activating chemicals relevant to the chemical initiating differentiated cells function and with the aid of incoming sustenance from the mother additional cells are produced eventually forming an organ etc. These incoming specific chemicals relevant to the differentiated cell of fixed functions results in the development of an outer membrane with receptors peculiar to these chemicals. This process revolves around male and female being exposed to similar environments and therefore evolving compatible genes etc. When the process of fertilization is repeated and the offspring is female the gamete process occurs once again and the evolutionary process continues uninterrupted as it has done since the beginning of the Eukaryotic life process.

Fertilization occurs when the sperm, equating to the original second invading prokaryotic species, with a single complement of DNA chromosomes, invades the female produced gamete egg as it proceeds down the fallopian tube and creates a situation where it is exposed to propionic acid molecules, a high energy bio-chemicals that can ingress into the cells cytoplasm and kick start the mitochondria process of energy release from glucose molecules.

Now being subjected to a plentiful supply of energy via the sucrose, day and night the eukaryotic cell is now equating to the energy susceptible stage that in its original historical condition of a prokaryote species led to it splitting into daughter cells.

On fertilization the germ cell produces a batch of cells (The Embryo) one of which when activated forms the Ovum organ. Gametes or egg cells, with single DNA chromosomes are then produced within the organ (The embryo consists mostly of uncommitted pluripotent cells, i.e. the DNA of all somatic cells is the same and the cells are capable of performing any function of the species given the application of the initiating environmental energy pulses which come via one of the randomly allocated chemical packages from the egg yolk that has been upgraded (if necessary) by the influence of any persistent environmental changes the female has experienced in her lifetime.

This system, along with a similar affect on the organ producing the sperm of the male resulting in the sperm being upgraded achieves ongoing evolution. The necessity for the genes etc of both parents to be compatible (similar) has developed, as corresponding genes etc of the paired DNA chromosomes, should normally have evolved in response to similar environmental influences and therefore complement one and other and may interact with genes etc originating from the other partners complementing chromosomes thereby making the possibility of dysfunctional performance a real threat if significant differences have occurred due to environmental diversification experienced by the preceding generations.

The designating chemicals come into action, activated by the application of heat energy, and they initiate the activity of individual embryo somatic cells (all initially the same and uncommitted) along the pre-conceived evolutionary path when they prompt the activity of the DNA of the cells in order of evolutionary priority as they respond. The chemicals acting as a conduit for environmental energy affects are energy activated to break down releasing the specific energy impulse influences to the cells DNA. The initial energy (heat) activation supports and develops the embryo cells as they are guided by these initial differentiating influences to produce further cells, along the same timetable and pathway as experienced during evolution, until outside influences take over (Firstly as the offspring commences development from the zygote (Germ) cell a process "meiosis" forming tens of thousands of the females haploid cells (Gamets) for storage for future use occurs as part of the development of the embryo when cleavage, a process where many pluripotent somatic cells randomly match up with environmental derived chemicals from the egg and on the application of heat energy form the blastula with differentiated cells and where gastrulation has taken over and then the placenta process and when born the food and external environment take over in supplying the life inducing energy and activation affects).

For the eukaryote species development, internal cells needed to be accessed not only by the initial activating and operating energy (heat, mostly delivered by sucrose molecules) but also by the environmental energy involved in the differential responses thereby triggering the responses of the DNA via the genes etc as the characteristics evolved over the generations are activated. For the complete range of these species different systems have evolved depending on their historical environmental exposure, however they all achieve the same objective of development and maintenance of internal cells. For eukaryotic species this has been resolved by the system evolved, complementing the original system of direct exposure to the environment. The system is;

A combination of direct exposure to environmental energy and evolution of a secondary system of bio-chemicals supported by the intake of sustenance (Should be from same habitat as the specimen) initiating a role in the cell of a response of producing bio-chemicals capable of activating further evolved internal differentiated cells

As the species environment persistently changed the evolution of further cells to perform adjusted and complementary functions occurred. The changes involved with the nucleotides of the DNA can only physically involve replacement and adjustment of a very small number of these at a time and as a consequence changes to the species environment that it can endure can only be slow and restricted.

Considering the situation of the evolution and functioning of internal organs etc associated with eukaryotic mammal species as the environment changes the following applies.

Bio-chemicals produced by and controlled by the cells of the digestive system are developed from foodstuffs that are produced directly and/or indirectly in response to environmental energy and as such they are conduits for the affects of the energy responsible for their existence. The energy pulses released when they are chemically broken down reflect the environment they were raised in and the response induced from the relevant cell DNA provides the necessary support, maintenance and growth for the dependents species survival. When a persistent change of the release of the pulsating energy environmental effect from the sustenance occurs the specific cells of the organ etc, about to be duplicated that normally replace damaged or provide extra cells to carry out the existing organ functions will be evolved by an adjustment to the DNA genes etc. resulting in cells able to counter the affects of this secondary environment change. This change will eventually possibly result in adjusted, complementing organ, gland etc or even limb, feathers etc to ensure the survival of the species.

The about to propagate cells, (fatigued, deteriorating cells) with their DNA nucleotide pairs splitting and hence duplicating having had their DNA (Genes etc.) adjusted by this hit and miss chemical replacement of the involved nucleotide pairs due to the persistent divergent energy input effect of the environment, results in cells capable of producing a modified chemical output (proteins, enzymes etc) (evolution of a supporting organ or adaptation of the relevant organ) to complete the process and it is a secondary means of adapting to an energy change, directly associated with the environment.

For these specific internal cells the affect of the modified chemicals i.e. proteins etc from the sustenance raised in the persistently changed environment is a conduit for the environmental energy whilst the supplies of basic incoming chemicals are utilized for the common tasks such as production of cell membrane, DNA molecules and the cells organelles etc to establish new cells (organs, glands etc), with appropriately modified RM and genes. When the species is exposed to an environment with changed elements in significant amounts, the species may or may not evolve to utilize these elements enabling survival. As the pulsing energy effect on the DNA, released from the chemicals, initiates continuing reproduction of daughter cells, the number of cells reaches a balance equalizing the deterioration and destruction of similar cells as a result of wear and tear on them, culminating in stability of the organ (growth stopped) resulting in the balanced provision of the supporting characteristic that provides the species modification relevant to survival in this changed environment effect.

Survival of the species is due to the environment manipulating the DNA, converting it to develop cells that suit the circumstances and in the instance of the cells being directly exposed to the environment i.e. skin cells, a process where the direct pulsating energy of the habitat is utilized to activate a response from the DNA i.e. the production of vitamins has evolved. These cells however have evolved in response to the primary input being supported by a secondary input of energy from sustenance ingestion resulting in them being more efficient.

Brain cells have also evolved to process the incoming information in a suitable but different manner, however the end affect is the same, pulsating energy from the environment via the evolved sensors induce an effect from the cells DNA molecules due to an harmonic response from the relevant nucleotide pairs, which results in an output, as it does for all cells, attempting to control any further potential damage and ensuring survival of the DNA and the specimen as they respond to incoming environmental chaotic information changes.

This process is not perfect and never can be, otherwise the DNA's response would be completely compatible with the environment and it would cease to function. (React).

The process as described above establishes that as a specimen goes through its life cycle the reactive section of the DNA chromosomes of the cell of an organ is adapted (evolved) during its lifetime, to cope with a changed environment e.g. if an applied energy impulse responsible for the development and well being of a kidney persistently deviates it can result in the DNA of the relevant kidney cells being adapted to cope with the environment. (This indicates that DNA does not necessarily remain exactly uniform throughout the somatic cells of a specimen during its lifetime, as the influences responsible for evolving the DNA of these cells are not accessing other cells, other than the gamet and sperm cells involved with the reproduction process).

A further example of this is the brain cells where the "Junk" section of the specific cells cannot cope with the reception of increased designated input and therefore under this stress enzymes are produced that facilitate the extension of the nucleotide pairs of the relevant cells during the specimens lifetime. The result is different capacities of the "junk" DNA may exist in brain cells, where necessary, however the Gamete cells and the organ responsible for producing the sperm cells receive all of the influences extending the Junk DNA where necessary with all of the evolutionary information. The consequence of this is that at the start of the offspring's life the entire DNA throughout the Embryo cells has the same Junk DNA, allowing the process of random selection of these cells for any given purpose to proceed.

When the change of environment entails a reversion to a previously existing environmental affect in the "sex linked" chromosome pair a further gene(s) is developed in the non-sex linked chromosome and the gene it is overriding is blanketed by a methyl effect. The process occurs in this manner, as once established the genes that are damaged nucleotide pairs, cannot be eliminated in the conception process. The new gene, known as a recessive gene, along with a similar gene from the partner results in evolved change.

It is a more complex system than just one energy impulse being dealt with per cell and some environmental changes may occasion sophisticated reactions.

As species have evolved a complex integrated system has developed where parallel developments occur to cope with a change of environment e.g. pancreatic cells to produce Insulin to control the level of blood sugar access to cells, along with liver cells controlling the release of appropriate amounts of sugar into the blood stream.

Upstream cells cannot normally be produced from a cell that has developed later than it in the process of evolution as the membrane of the cell does not have appropriate receptors and excludes the appropriate chemicals to the cytoplasm and this is tantamount to the DNA genes being committed. However as each and every somatic cell nearly always has exactly the same DNA, intervention prior to the consolidation of the membrane and provision of the relevant environmental energy impacts will allow any cell to be produced from any existing cell. Skin cells utilize a different process of exposure; however it is apparent they also could be adapted under the right circumstances.

The DNA of each cell of an animal species is packaged in distinctive multiple numbers of pairs of chromosomes consisting of two lengths of DNA with a joining feature. These pairs consist of a chromosome from each parent with the DNA of each having a similar layout with the exception of the chromosome pair governing the sex characteristic

(X, X) or (X, Y). Each pair of chromosomes is encased in a layer (Chromatin) of chemicals, mRNA etc and coating of protein pieces known as the epigenetic effect. The DNA also has areas coated in Methyl, a chemical that bonds to acids i.e. the nucleotides, changing their resonance frequency and rendering them unresponsive when normal environmental energy vibrations affects are applied. This methyl tag is deposited on parts of the DNA, making it inactive and therefore enabling the alternative gene(s) to the gene(s) blanked out to be temporarily dominant until the methyl is removed. An example of this is the sex characteristics that develop in puberty.

The DNA of these cells must be accessible to all of the incoming environmental energies as each of the four possible strands of the DNA of the uniting pairs of chromosomes must be completely available for possible selection during union i.e. male or female and therefore the tag of methyl normally controlling unwanted reactions is shed. This process of methylation has evolved to control a required sex orientated reaction along with the inactivation of unwanted portions and foreign inclusions in the DNA. As the specimen develops and sex characteristics are scheduled to develop the appropriate methyl tagging removal process is activated. This process is a reaction resulting in the DNA chromosomes carrying both controlling sexual characteristic sections in the genes for evolutionary and reproduction purposes. (This process is similar to the process of controlling the differentiation of the lactose consumption-initiating gene of the E-Coli)

The methylation system has evolved, assisting in controlling differentiation when the development of the alternate sexual characteristics (sex linked) throughout the individual is activated. Absence of the methyl tagging in the gamete cells of the female and the sperm cells of male sex organs is based on the DNA chromosomes being available for total non restrictive evolutionary inputs and reproduction purposes, and when reinvoked in the cells throughout the developing embryo temporarily inactivates all of the unused sections (Genes that can be involved in the production of proteins enzymes, hormones etc resulting in characteristics that are not active at this stage.) of the DNA including that inserted by virus infection. This results in the inert virus contamination forming approx. 50% of the human DNA being propagated. The process allows all of the various environmental energy effects established over the species evolutionary history, including ongoing changes; to be represented by specifically formulated Proteins possessing the relevant energy affects when chemically reduced, to be included in the gamete cell. When the sperm, already updated in an evolutionary sense, due to the exposure of the male to persistently changed environmental effects over his lifetime has fertilized this cell, pluripotent embryo cells are produced with their DNA chromosomes having one half of the complement of gene nucleotide pairs installed, strained beyond their elastic limit.

When exposed to the resonating effect of the heat energy, the DNA chromosomes nucleotide pairs of the embryo cells vibrate rapidly, resulting in the installation of the complete array of the genes etc. with the complete environmental information installed that can be accessed selectively when the appropriate energy pulses are applied i.e. the cells are pluripotent at this stage before the energy pulses are applied i.e. the necessary information for establishing the physical makeup and adapting to any evolutionary information is downloaded to the new pluripotent cell, with its DNA being at first reproduced in an inert chemical form and then during activation, imprinted with the gene information and then further differentially activated to cope with the environment resulting in the establishment of the specimen.

After the reproduction act the and the embryonic cells are being formed with the sex established, the developing DNA is vibrated and activated by the heat energy pulses and the information processing capability (genes etc) installed (Imprinted) and then the differentiated property of the cell is installed by the heat activated bio-chemicals releasing environmental energy pulses of a frequency where they access the appropriate genes etc with the forming membrane of the developing differentiated cell establishing the appropriate receptors, enabling continuing reactions of the developing cells of this particular role to be established. The various differentiated cells then control all of the characteristics of the developing species. The methylation, telomeres and "epigenetic" installation effects occur simultaneous with this process. The methylation effect is installed to the virus contaminated DNA, along with the involved genes that are required to be inactive, temporarily rendering them inoperable.

NOTE

1. Every cell produced or has been produced has DNA organic molecules as its reacting entity and as such the DNA has the normally available elements, as described above, incorporated into its makeup. When DNA is first formed it is activated resulting in the inclusion of genes etc and evolution if established as a result of environment as necessary to survive and this is a process that occurred in a very similar manner in the beginning of the life process, in fact the basic process does not vary, but over the billions of years of its existence, in reaction to exposure of the DNA to many and varied environmental exposures has resulted in outputs developing into millions of species, many with sophisticated secondary processes, but always based on the same initial characteristics of the DNA.

2. The truth is all species are beings of the environment and their characteristics, (All characteristics) are developed as a result of the property of the DNA molecule being susceptible to the resonating effects of the incoming energy pulses representative of the environment. Genes are installed as a result of the damaging effect of these incoming pulsing energy effects from the environment or changing environment and a chemical reaction that extends the molecule as the destructive damage is repaired. They are a permanent effect lodged in the DNA molecules thus ensuring its ongoing survival as a record (Memory) in descendants as it serves to provide protection against the ongoing damaging effects of these environmental energy pulses as it (they) initiate a cell reaction output (the species) leading to its survival in this particular environment as it reacts and evolves.

3. The scientific process of Life and Evolution is ongoing and had a beginning, as it must. A true evaluation of this process demonstrates that such concepts as Creationism, Intelligent design and Darwin's Theory of Evolution are not viable concepts.

When the species environment changes so do the incoming energy pulses. (All aspects of a species environment always access the DNA nucleotides by an energy impact i.e. sound, light, heat, odours, scenery, sustenance etc are always delivered to the DNA in the form of energy pulses either by primary or secondary means).

Note

1. Ongoing miniscular changes over a period of a billion years have
    led to apparent extremely complex detailed survival adaptations;
    however the basic life process is always the same, development
    of the DNA genes etc to provide the damage control against
    the effects of the incoming environmental energy on the DNA
    nucleotides. Each and every cell of every species follows the
    same basic process as described above, with the cells life process
    activated by the relevant incoming environmental energy input.

Each and every characteristic (Evolved as a reaction that contains
the damage caused by persistent incoming environmental energy and
this maybe physical or mental) of each and every species is subject to this
system. The DNA never varies in its basic properties, however depending
on the particular environmental influences the species has and is being
subjected too, varied and particular output changes occur. This leads to
a situation where once a diverging situation has been triggered it is built
upon complementing the first change and this establishes a trend where
although two or more species appear to inhabit similar environments
they have divergent characteristics. Additionally to ensure survival of
the DNA a system has evolved allowing reactions. (Always based on the
original properties of the DNA) to different aspects i.e. increased, of
the incoming environmental energy affecting the species, as it possibly
develops more complex divergent characteristics. The system is one of
degree. In the human species in particular the consciousness capacity
is a multiplication of the number of glial cells present (billions), by the
evolved capacity of the awareness trait of the DNA, initially present in
the conception (Zygote, germ) cell of the specimen. As information is
loaded to this evolved capacity by the environmental stimuli (energy),
during the lifetime of the specimen, the performance changes and if
the loading of specific, new and changed information(s) is persistent
enough prior to reproduction, changes occur in the physical state of
the parents relevant dedicated DNA and the offspring at conception
is adjusted (evolved) as previously described, to the environmental
affects responsible for this change. The system is similar for all mobile
specimens but the results are varied according to the historical exposure
of its ancestors and ongoing lifetime exposure i.e. nurture is responsible

for nature happening. Various species have evolved utilizing different systems, however the final result is always the same; the ongoing survival of their DNA.

In the more complex species as previously described, sustenance is not only used to provide physical building requirements and energy for the species but also a system has evolved for breaking down chemicals within the cells, with the commensurate release of energy pulses to a frequency that then have the same evolutionary effects as direct exposure to environmental energy on the DNA. Under normal circumstances the food sustenance will have grown in the species environment and reflect the mostly similar environmental effects applying to the species when the energy released reacts with the DNA and establishes a response. In other words one of the functions of the food is to act as a conduit for the energy pulses of the environment. The situation then augurs badly for the human species when, parents from different environments having consumed foodstuffs produced in significantly different environments may result in problems of incompatible DNA being present in offspring. Human cultures who have a history of evolving in differing environments and therefore possibly possessing incompatible genes etc demonstrate such problems as diabetes type one, skin cancer, allergies, mental problems etc. (This conclusion may be checked out statistically against stable populations i.e. China, India etc. versus emigrational conceived populations i.e. Australia, USA, NZ etc. Further the inheritance of characteristics of appearance is extremely diverse in these populations as against the stable populations and this situation occurs with all characteristics) (Ref 4). Genetically modified foodstuffs can also pose problems, as can different foodstuffs from different environments leading to allergies etc. A further problem area is that as intakes similar to sugar, such as alcohol and drugs are dealt within a similar as manner, they disrupt the process causing basic problems associated with evolutionary and inheritance processes.

So-called genetic problems (They are not, they are DNA problems, which is the dominant entity and genes are secondary to it) can then be examined and the causes established leading to possible treatments and cures. Some of these problems result in cancer, diabetes, and dementia, M.S. etc.

As an example, the cause of the problem of cancer is described below

Any foreign substance ingested or an oxidant or foreign energy pulse may cause damage if it penetrates a cell ready to commence division and affects the gene, disrupting the evolved coding arrangement controlling the makeup of the formulae of this chromosome protective device (Chromatin, "Epigenetic" protein). The coding of the gene being disrupted then controls the makeup of the chromatin formula that is no longer able to provide the necessary protection against the normal environmental energy pulses and they gain access and further disrupt this protective characteristic of the chromatin as the gene coding is damaged further. As the energy pulses gain greater access they initiate a reaction from the RM (Regulatory memory) that controls the rate of propagation of the cell with the result the cell propagation runs amok and as it is a cell dedicated to the production of cells it can cause the eruption of cancer as the cell production system goes out of control creating dysfunctional proteins etc resulting in aneuploidy of the chromosomes, with the dysfunctional cells still able to divide and create a parasitical organ. There is an effect, "Apoptosis" that will normally cause the damaged cell to self destruct however if the cell is ageing or there is a fault with the involved gene(s) and the RM guiding the production of this lethal protein and the system is rendered ineffective the dividing cells continue in their task, producing daughter cells that go on producing further disrupted cells. The same problem may erupt with certain virus infections and in the case of an adjacent pre-existing mutation in the DNA, with the incoming environmental energy varying, or a break down in the energy access system possibly caused by a disruption from the epigenetic coating or stray energy pulses of a non environment nature the DNA of a dividing cell may carry damage that eventually causes it to erupt in response to an energy effect as the Apoptosis affect fails (e.g. inherited susceptibility to breast cancer).

In summary life can be described as the reaction of unstable organic molecules of DNA, incorporating radio-active elements, to the continually changing effects of their environment, where they are activated at all times to be compensating (resisting) any potential damage from the incoming varying energy pulsations, and thus survive. Each change of frequency applied to the DNA due to a chaotically fluctuating environment elicits a response from the DNA of a protein etc or mental output and therefore each evolved characteristic represents survival in environmental energy frequencies, past or present. Where there is an absence of environment energy change, the molecules are non responsive and consequently "life" non-existent.

Evolution is the updating of the DNA, as it is revised to cope with the changed environment effects where the change is persistent over time, leading to the species (DNA) changing (evolving) and becoming suited to the new conditions.

The ability to think is a development from the original Regulatory Memory characteristic of directing the reactions of the DNA to limit damage and ensure survival in the environmental energy and the number of cells evolved to cope with the incoming information from the environment magnifies this ability. The reaction is always initiated by an external input to a relevant brain function that then sets of a search program involving logic and memory access.

In relation to the human species as in all mobile species, behavioural traits are present as result of the complex thinking ability, and therefore are subject to the inheritance process. As a result of our greater mental capacity our species range of behavioural characteristics varies more than any others do. This situation may involve criminal behaviour, violent behaviour and in fact the whole range of human behaviour and is therefore extremely difficult to control. This then highlights the attempted method of control the human race has adopted for aberrant behaviour of incarceration as a punishment. The individual involved may not be in control and therefore the control system adopted should perhaps be the protection of others from further problems.

The process of evolution and the cause of the life phenomena are common to every life form beginning at its inception and ongoing during the species lifetime. The concept of the life characteristic, as with any other characteristic can vary from species to species depending on the historical exposure to the environment of a species ancestry.

All cells of all species are semi independent (or in some cases physically independent) and life is the accumulation of the integrated effects of the reactions (awareness) of all the integrated cells, governed by their evolution i.e. for mobile creatures examples of various aspects of life are present in the supporting organs, driven by the energy pulses (stimuli) of the environment responsible for their evolution and consequently "Life" is the ongoing accumulated effects of the reactions of the DNA as it strives for survival (compatibility) in the damaging (chaotic) effects of its environment. This situation is demonstrated by the ability of scientists to be able to manipulate cells, imagined to be "stem" cells but are cells with DNA that having been exposed to the riguors of its incoming energy pulsations suffers fatigue of the nucleotide pairs and splits apart, then a chemical reaction aided by an enzyme action rebuilds the two halves of DNA and the cells are duplicated. Given the correct application of energy effects, prior to the membrane with its differentiating membrane performing the receptor function forming, this renewed (Pluripotent) cell can be induced to develop into any cell.

Life then in some respects can be regarded as an illusion and the original DNA of the embryonic cells of the eukaryotic species contains all of the directives and potential characteristics from the parents including being poised, as the cells are activated by a heat energy input and for its genes etc to react to a variety of additional environmental energy effects delivered by chemical means, a process evolved over the lives of the generations of its ancestors. This system of evolution revolves around any change of persistent environmental information that generations of parents have experienced up till the time of conception. This enables the offspring to survive in the ever changing and updating environmental energy impulses and thus the environment. Members of a species are normally exposed to a very similar environment and should be, thus ensuring similar and compatible characteristics evolve.

References:

1. Pjotr Garjajev and team of Russian biologists (refer http://www.spaceandmotion.com/Evolution-Biology-Wave-Genetics.htm
2. http://www.newscientist.com/article/dn3548-electrifying-claims-for-dna-are-dashed.htm
3. One Face, One Neuron Dianne Martindale Scientific American Oct 2005
4. Environmental Cross Breeding Extract from Melbourne Herald Sun newspaper.
5. Making Memories Stick R. Douglas Fields Scientific American. Feb 2005 (Copy Attached)

6. The Hidden Brain. R. Douglas Fields. Scientific American Mind. May/June 2011 (Copy Attached)
7. Copy of Email re cause of MS.
8. Protein, Its role in disease.
9. Epigenetics, Evolution

For more see my website http://evolgenmutsuic.com

Note:

The order of appearance of a specimen's characteristics parallel the order of the evolution of its characteristics and when a new characteristic evolves this characteristic appears last as the specimen develops.

FUNCTIONING OF CELLS

INTRODUCTION

This paper examines the primary process of how cells function and then delves into the process of cell function in multi cell Eukaryotic species including how mammals became symbiotic species. It further demonstrates there is a basic process underlying the function of all their types of cells i.e. differentiated that in fact applies to the cells of all mammal species and it relates to the reaction of the DNA molecules and establishment of genes by the effects of the incoming environmental energy pulses associated with its environment.

It does not delve into the myriads of secondary detailed effects evolved over 100's of millions of years where the DNA molecule initiates, as a reaction to these environmental effects, outputs from the cells resulting in their survival and hence the species in its environment. These reactions of the DNA molecules of the mammal species cells initiates the incorporation of many readily available elements such as Ca, Fe, Mg etc. into their outputs providing their integrated physical and mental characteristics leading to their survival in their environment.

Concentrating on the primary aspect of a cell function highlights just how misleading getting involved with the secondary aspects can be when trying to understand the causes of genetic problems such as cancer, diabetes, MS, Autism etc.

It also emphasis the necessity for a better understanding of the sciences of physics phenomena and engineering principles, not at present apparent in basic biological research.

Further it emphasis there are several beliefs held in Biological dogma that are illogical, unscientific and complete blockers to the pursuit of the truth in Biological/Genetic research.

These beliefs are

a). Species evolve to survive of their own accord in a changing environment. This is an impossible concept as they would not exist in the first place; in fact it voids the whole concept of the possibility of life.

b). Cells propagate and react spontaneously. Once again, not possible a basic law of physics states "Every effect has a cause".

If bemused by the hypothesis then consider the following,

1. Every species is compatible with its environment

2. Every cell has DNA molecule(s) that is its only reactive entity.

3. The conception of any species and its development follows its path of evolution in a constricted time frame e.g. in the mammal species, the DNA molecules form in a damp energy controlled environment where the appropriate elements are available.

4. The Prokaryotic cells involved in conception of Eukaryotic species of both the female and male have chromosomes of DNA that are single stranded, whilst the female cell has the mitochondria cell incorporated.

5. The role of gut bacteria is established revealing mammals are a symbiotic species. The gut bacteria releases propionic acid molecules, a high energy bio molecule that can have isomer forms as the gut bacteria is evolved when exposed to a variation of incoming energy pulses such as anti-biotics when being chemically reduced within the gut.

6. The male cell (sperm) penetrates the female cell, through its outer membrane, allowing the propionic acid molecules, present in the mothers blood flow, entry that activates (kick start) the mitochondria cell, consequently activating the DNA molecule by the effects of continuing energy impulses now available from the incoming sugar molecules being chemically reduced, resulting in the genes being installed and ready for the differentiating activation to take place.

Function of a Cell (Eukaryotic, Mammal)

Examining the process of life in general DNA is the reactive entity of a cell. It is the only reactive entity in all cells of every species. It consists of four nucleotide acids Adenine (A) Guanine (G) Thiamine (T) Cytosine (C) plus sugar molecules with which these acids are chemically integrated. Phosphate atoms bond the sugar molecules into the one sided rail like structure of the RNA molecule. Two compatible molecules of RNA then bond together forming a spiral ladder like DNA molecule with the compatible acids (CG) and (AT) being bonded together at their protrusions by hydrogen atoms, forming the "rungs".

In the beginning in a soupy swamp minimal lengths of RNA molecules formed consisting of chemical molecules of the four natural occurring nucleotide acids and sugar. These RNA molecules consisted of the readily available elements (C, H, O, N and P) making up the acid molecules in indiscriminate arrangements (half sides of DNA molecules). Eventually with the passing of time complementary pairs of the acid protrusions associated with the RNA molecules bonded via hydrogen atoms forming the complex molecule DNA.

Eventually the DNA molecule with the nucleotide pair arrangements were such they were in a condition where a harmonic vibrations occurred as a result of the incoming energy pulses emanating from the environment. (All effects of any environment on a species are received by the DNA molecules by way of relevant energy pulses to which they react initiating outputs from the cell related to their survival in this environment and when there is a slight persistent change of the environment the incoming related changed environmental energy pulses damage the nucleotide pairs and a hit and miss enzyme assisted chemical replacement and nucleotide pairs addition occurs resulting in a modified arrangement of nucleotide pairs that are once again in a harmonic state with and compatible with this changed environment effect with an output from the cell initiated by the appropriate gene(s) reaction tending to maintain this state, leading to the DNA's survival and hence the evolving specimen).

Initially the rung like pairs of nucleotides of the DNA vibrated (A DNA molecule is subject to vibration, Prof Pjotr Garjajev, Russian biologist) when this arrangement of nucleotide pairs was exposed to the impact of pulsating energy (Heat energy from the environment. Emitted energy beams are always constituted of photons travelling in a helix with different frequencies and amplitudes and on impact with a substance energy is transferred in pulses) (The phenomenon of Vibration and Resonance is the most common physical phenomenon in the universe). When a persistent change of energy pulse of a fixed frequency from a slightly changed environment, such as from a variation of a colour, the vibrating effect, when directly accessing DNA nucleotide pairs was sufficient to cause destructive damage to those now out of tune.

From the beginning of the life phenomenon as time moved on a protective protein enshrouding the DNA molecule and latter a membrane evolved (The cell), utilizing the available chemicals from the environment and those present when the nucleotides were broken down as the DNA molecule became adapted to survival in the developing environment. As this time dependent process moved on further developments evolved within the cell membrane resulting in enzyme assisted chemical reactions and ribosome structures that aided the production of proteins being established, with the chance of the DNA molecule surviving in the environment, becoming more positive.

The whole process was governed by a default hit and miss process of replacing and adding to the reacting nucleotide pairs (The Regulatory Memory, RM) until they were once again in harmonic tune with incoming relevant energy pulsing effects of the environment. The enzyme aided chemical process occurred in a hit and miss fashion as no plan as such existed to replace the nucleotide pairs, but a developing "coded" control process of the DNA nucleotide pairs eventually resulted in the cell producing a revised protein sheath (Chromatin, a protein shield, wrapped around the DNA chromosomes in the nucleus of the cell that eventually results in a negating effect on the normal incoming environmental energy pulses, where the tendency to cause further damage effects to the genes and RM of the DNA chromosome became controlled).

As further changed and/or additional incoming environmental energy pulses of a distinctively different frequency came into being, but not overwhelming different, breeching of this chromatin protein occurred, as it had not been evolved to provide the necessary protection against this energy frequency, nucleotide pairs of the DNA molecule more akin (Tuned) to the frequency reacted with the further development of RM nucleotide pairs and genes (coded nucleotide pairs) with further outputs from the cell of protein etc. being produced that added to the survival characteristics of the cell in this environment. The phenomenon of life and evolution was then underway.

The outputs of the cells were dependent on the responses of the genes leading to activation of the cells and these genes came into being as follows, as with the changed energy input several of the existing reacting nucleotide pairs become no longer harmonically tuned and being redundant to those nucleotide pairs that were respondent to the particular incoming energy pulse (Tuned) and although not destroyed by the energy vibration event, were strained beyond their elastic limit and no longer tended to vibrate. These strained nucleotide pairs, the gene, now basically represented a record (memory) of the previously prevailing environment effect (Developed over generations of exposure) as their pairing had been initially coded (arranged) as a result of the environment effect they had been exposed to and with the energy event persisting in a slightly changed format (see latter for detailed explanation) it initiated a hit and miss mechanistic chemical reaction replacement process from the cell via the gene that eventually led to output of proteins etc. with the characteristic of the ability to negate and survive in the effects of this environment.

With the advent of multi cell species, over 100's of generations and exposure to an ever increasing complex environment resulting in regular, freely available energy sources genes were further developed in response and with the extension of the DNA molecules with increased numbers of genes the increased numbers of outputs eventually resulted in diversification of the cells (differentiation) with the resulting output from the controlling cells leading to the conscious life effect due to the phenomenon of the relevant genes (memories, records) and "Junk" DNA extension being adjusted by the environmental energy pulses, activating and evolving the genes, that were being deflected off nearby sustenance. This meant an extension of the brain cell numbers related to the evolving sensory perceptive sensors necessary to cope with the mobility and control of limbs that were evolving to deliver the necessary outputs of the reaction of the cell's DNA molecules to these input (Leading to survival) that were coordinated on delivery to the initial brain cells (Neurological) whose output then had the necessary coordinated outputs to logically control the developing limbs.

The initial energy, gene, ribosome process resulted in a protein (Ref 1.) being produced where it assisted in the survival of the DNA chromosome helping to counter the local damaging effect of the energy (Scab like) and as the nucleotide pairs are now being subjected to the persistent, but muffled energy pulses this localized output of protein is maintained i.e. the DNA molecules (RM) being exposed to the energy pulses activating the differentiated cell gene(s) results in a reaction with an output of protein resulting in its survival. (In Eukaryotic species these energy inputs may be provided by primary or secondary means where a carbon based bio molecule, that is a conduit for the environmental energy and therefore is representative of the environment it was produced in, is chemically reduced within the cell, releasing controlled energy pulses specific to that bio- molecule.).

The initial localized chromatin protein protecting the DNA chromosomes consists of a protective or muffling protein around the chromosomes with the "Histone" effect (Encl 3.) being a clump of protein, forming beneath the DNA molecule section and forcing the chromosomes RM's and genes clear, due to their writhing effect as they vibrate, leaving the genes free to initiate a reaction to the incoming energy pulses. This effect of exposing the Regulatory Memory (RM) nucleotide pairs that are the activated component reacting to the relevant environmental energy pulses results in the appropriate expression of RNA to be initiated by the gene as it is "switched on" by a generated energy signal from the RM. (See latter paragraph). The Histones appear at several locations along the nucleosomes where the genes are activated in responding when involved in the output of the differentiated cell. These histones are present, at common genes in the various differentiated cells as well as at the genes specific to the differentiated cell. This process of developing genes results in the addition of or change to the original cells output, as the DNA and hence the species characteristics are evolved.

The continuation of the energy effects on the exposed genes results in the ongoing production of protective chromatin proteins that form an amalgam of proteins providing a protective muffling effect for the unaffected DNA inactive genes, nucleotide pairs etc. This amalgam is the chromatin providing the so called Epigenetic effect, as in the absence of a knowledgeable explanation by others; the chromatin is believed to be coded as the ammonic acid molecules arrangement in the protein varies from one type of differentiated cell to the other. Apparent inheritable variations of the chromatin associated with genetic problems can also be present as a DNA mutation can result in the production of a changed chromatin protein and these changes have been assumed, for want of better knowledge, to be the driving force of a disease and not a symptom as it actually is. The so called Epigenetic effect is thus a symptom of a genetic problem, not the driving entity of the problem!

This is the initial part of the process resulting in the DNA and hence the cell reacting and providing self-induced protection with survival occurring in an environment. It is also the cause of the variations present in the DNA of specimens of a species and for the lengthening of the DNA strand with many additional genes and RM sections as it is subjected to, over the generations, slight gradual changes of the species environment. This then is evolution with the process of inheritance tending to stabilize the evolved characteristics within a group exposed to a similar environment.

For the DNA molecule to survive in the damaging effects of the environmental energy pulses and therefore the environment protection must be available, and development of relevant proteins occur in a controlled hit and miss process. The change of the effects of the environment therefore must result in gradual and minimal change of the energy inputs. This is assisted due to the protective protein (Chromatin, Epigenetic effect) surrounding the DNA that is formed chemically as a result of the incoming energy activating the RM associated with a gene that has evolved, initiating the chromatin protein protective response from the cell that muffles DNA molecule from the effects of the incoming naturally present environmental heat energy and that provided by the sugar and the particular Bio chemicals associated with the differentiated cell as energy pulses are released as these molecules are chemically reduced within the cell, thereby extending the life of the DNA molecule and hence its cell.

The chromosome molecule is split apart as a result of eventual material fatigue of the hydrogen joints as the reduced energy vibrating effect accesses the nucleotide pairs. (On splitting a chemical process initiated by existing chemical enzymes etc. produced within the cell restores the two half sides of the now existing RNA molecules forming two sets of DNA and hence two propagated cells (cells are NOT spontaneously formed. It is impossible)). Without this effect the DNA molecule would be rapidly damaged beyond repair or rendered dysfunctional making reactive responses from the cells and therefore life impossible. The nucleotide pairs of the DNA molecule being sensitive to the continuing energy pulses and reacting has a continuing capacity to negate the effects of the chaotically continuing environmental energy pulses with fixed frequencies and amplitudes as they are applied resulting in chaotic reactions (life effect).

When a slight persistent environmental change is experienced the associated divergent energy pulses rapidly damage several RM nucleotide pairs as they are no longer resonating harmonically and the hit and miss process as described above results in the nucleotide pairs of the RM and associated gene being readjusted, negating the damaging effects of this environment change as the existing protective chromatin protein is adapted by the newly coded initiating process of the gene nucleotide pairs resulting in revised amino acid molecules, restoring the muffling effect and leading to an extended life (variable for differentiated cells) for the cell. Additionally as a result of this change of energy input further changes to the DNA chromosomes nucleotide pairs occur (The "coded" phenomena) resulting in further changed integrated outputs from the affected cells of proteins etc. being produced leading to the specimens and the species being further adapted for survival in the changed environment. All the DNA molecule changes induced into specific differentiated cells are reproduced in the cells of the males sperm producing organ (Its cells are not differentiated) and as a result the sperm carries all changes in the male DNA chromosomes incurred up till the time of inception.

As the female has a bank of egg cells (Gametes) from birth they are protected from and are inert to any environmental evolutionary effects before fertilization (If this was not so the energy effect required for installation would need to be active, with the gametes being activated en-masse). As she, and her cells, have also been subjected to the effects of environmental change, if any, the bio- chemicals that she produces that activate her differentiated cells are updated due to relevant evolutionary gene changes and when the uncommitted propagating cells are activated by the energy pulses emitted by the bio chemicals in a controlled chemical reduction in the cells cytoplasm the relevant energy pulses install the evolutionary changes in the RM and genes of the differentiated cell's DNA chromosomes.

The equivalent male chromosome is also subjected to this evolutionary change effect emanating from the female and if changes have already been wrought on his genes as a result of the effects of a different environment (Habitat) the above effect would lead to dysfunctional cell outputs and does when occurring. The DNA genes and RMs of male and female origin should be evolved in similar environmental habitats. The concept that genes should be divergent for the welfare of offspring is erroneous. (See Encl 2.)

Each cell in every species (with a very few exceptions, that have an alternative source of radioactivity providing a magnetic field), contains Potassium (K) an abundant element, in the cytoplasm of Eukaryotic species and within the cell membrane of Prokaryotic species, whilst the radioactive isotope of carbon (C14) is also always present. The Potassium invariable contains a radioactive isotope (K40) along with C14 that provides a surrounding magnetic field to the DNA chromosomes and on activation of a RM group of nucleotide pairs as they oscillate through this magnetic field an electrical current is generated in them. A surrounding energy force field is induced around the conducting nucleotide pairs by the generated electric current, and it represents the damaging potential of this specific environmental energy, as this field rises and falls in response to the incoming specific effect of the environment. (The electric current associated with the force field activating the differentiated gene(s) of the cell is restricted to the particular RM nucleotide pairs by being exposed to energy pulses with their relevant frequency response characteristics, by a property known as Quantum Tunneling. The cell membrane receptors restrict access to all but the appropriate chemicals and therefore associated energy pulses). The force field identifies (switches) the gene(s) associated with this environmental effect and they activate a damage control process via RNA with proteins etc. being produced by the cell that counter the damage potential of the incoming energy pulse associated with the particular environmental effect. The protein accurately reflects the properties required to ensure the DNA's and hence the species survival in this environment (This is a survival characteristic evolved over 100's of millions of years). In this respect the chromatin protein evolved, forming the so called "Epigenetic effect" filters the incoming energy effect normally present thus significantly negating the pulsing stress on the DNA nucleotide pairs thereby reducing the destructive fatigue effect on the rung like nucleotide pairs and increasing the life span of the DNA chromosomes and thus the length of time before the host cell propagates. The cells do not and cannot spontaneously propagate. (Law of Physics "For every effect there is a cause")

Note

The protein constituting the Chromatin "Epigenetic" effect is evolved and produced, as is all other protein etc. by the cell in response to initiation by the gene, and counters the incoming energy pulses commensurate with the role of the differentiated cell leading to survival. As such it is never 100% efficient as the cell would no longer function and life would not exist. The protein has been evolved muffling specific energy pulse encountered in the species environment,(Ref 1.) however there is no counter for aberrant pulses not normal to the environment and these can do significant damage to the DNA nucleotide pairs resulting in such problems as dysfunctional runaway propagation of cells i.e. cancer.

The concept of a protein (Chromatin or Epigenetic effect) produced with an apparent molecular difference due to the damaged DNA nucleotide pairs arrangements giving the impression, in the absence of the knowledge of how the energy input drives the cell, that it has the driving capacity to deliver a disrupted output from the cell in competition with the genes and further that its properties are inheritable is impossible and not logical. (The energy pulses, with divergent frequencies encountered over prolonged periods from man-made sources such as smart meters, wind turbines; cell phones etc. are typical of this problem and can do significant damage).

This phenomenon of coordinated differentiated cells involves the property of the cells to provide different outputs from the common DNA chromosomes when damage control responses are required to counter specific environmental effects. This "Differentiation" property has developed in Eukaryotic species as they have become more complex and many of their variety of cells have been evolved that rely on and produce secondary means of supplying bio- organic chemicals that when chemically reduced in the relevant evolved differentiated cell release a variety of fixed frequency energy pulses and therefore activating pulses for the appropriate genes. The operation of this process relies on the supply of the appropriate carbon based chemicals to the differentiated cell cytoplasm via receptors in the membrane of the cell.

This Eukaryotic process evolved due to the invasion of a prokaryote cell normally fuelled by the primary source of energy by the Mitochondria cell with its property of reducing an organic chemical molecule e.g. sugar and releasing controlled energy pulses.

This combined cell was then invaded by another prokaryote cell with a doubling up of the DNA molecules resulting in a more resilient cell.

This multi cell event, one of a number of other events, eventually resulted in the forerunners of mammal species, when in the presence of emerging lichen, with its propensity for producing and releasing sugar, a single cell bacteria fuelled by this sugar molecule released propionic acid molecules in the presence of the mitochondria invaded multi cell, allowing the propionic acid molecule to enter the cell membrane that was weakened by the invasion of the second cell, causing a rift to appear and the propionic acid molecule gains entry. The propionic acid molecule, extremely high in energy content having entered the cell cytoplasm resulted in the mitochondria cell being kick started whereupon sugar molecules were drawn in providing a continuous supply of energy. This resulted in the eukaryotic cell being provided with and activated 24 hours a day by an energy source and it was no longer driven by direct exposure to highly fluctuating energy levels.

Once these cells combined integrating characteristics was slowly evolved providing the controlling coordinating characteristics necessary as the multi celled species was further evolved with features resulting in its survival in the developing and changing environment. These initial cells when on multiplication into multi cells eventually evolved into the brains of various mammals, when as exposed to slightly varying environments, a series of different mammal species appeared

The propagation process of a eukaryotic cell is a progressive process initially starting with a chemical process of completing the duplication of the DNA chromosomes and when complete propionic acid molecules gain entry through the cells membrane activating the mitochondria cell releasing heat energy pulses and restoring the genes as the nucleotide pairs are vigorously vibrated (Aided and abetted by the fact one half of the involved nucleotide pairs are already strained beyond the elastic limit). At the completion of the genes being reinstalled and the propagated cells with their cytoplasm and membranes etc. being chemically restored in an enzyme assisted process it is in a receptive state ready to be activated (The pluripotent state). The normally present source of chemicals providing the energy pulses relevant to the differentiated cell then gain access to the pluripotent cell when the propionic acid molecule provided by the ever present gut bacteria aggressively penetrates the membranes and being high in bonding energy kick starts the mitochondria cells into action i.e. the energy provisioning cell, and in the presence of the differentiating chemicals the membrane develops appropriate receptors resulting in the multi cell being set to work.

All mammals are dependent on gut bacteria to provide this process and therefore have a symbiotic relationship with the bacteria, with the bacteria also depending on the mammal for survival.

Prokaryotic species have been evolved to have the property of changing reactions to different energy inputs as required, e.g. E Coli however it all results in the one phenomenon, survival of the DNA molecule as it reacts to the environmental energy input pulses and hence the survival of the species it supports.

With Eukaryotic species various types of cells have evolved, some reacting to direct environmental energy inputs for their activation, but their energy supply is by chemical means i.e. brain cells, whereas others have been evolved relying on direct chemical molecule environmental input i.e. kidney cells. The brain cells are utilized to store memories and action thoughts and to this end "Junk" DNA nucleotide pairs have evolved and constitute, in humans 93% of the approx. 3.2 billion pairs that exist in each cell. These Junk nucleotide pairs of the DNA of these cells record the incoming information when dedicated cells receive energy impulses from the evolved sensory organs when they deliver energy pulses suitable for causing harmonic responses to occur in the "Junk" nucleotide pairs of early exposed cells causing further energy pulses to be generated in a feeder process and sent on as are all the energy pulses emanating from the various senses where they are coordinated at the final cell in the memory route, the neuronal cell where the complete information is recorded, if the incoming information is consistently repeated (Encl 1.). The repeated effects of specific energy pulses on relevant nucleotide pairs (i.e. tuned) causes' mechanical strain to occur and with the material property of time based recovery results in sustained conscious memories of various durations when these distortions are re-accessed by this particular energy pulse. The strained condition of these nucleotide pairs, caused by the energy input from an environmental event then represents this event and on repetition is recognized as such (record or memory). Depending on the number of times repeated the information then becomes available as a function of the degree of strain and when identified results in a range of memory recall from slow to rapid recognition and when action is required resulting in survival. The installation of this information to the neurological cells is facilitated by a secondary reaction in the cells as specific dedicated protein is intermittently injected into the synapses between the cells as they pass specific frequency related information thus consolidating a memory. The synapse protein is initially established to pass only these specific energy pulses related to a sense and if the incoming information by way of energy pulses is not repeated it is demolished; however over time due to the controlled repeated exposure to the information being installed; the protein is consolidated, supporting the rapid recall characteristic.

The DNA, a chemical molecule does not respond to the environment but it reacts to the environmental effects and is adapted by it as it is variously affected by the different frequencies of the environmental energy impulses it is subjected to. The continual ongoing reactive adaptations of the DNA nucleotide molecules to various frequencies of environmental energy pulses developed over the history of the species (Evolution) and the ongoing reactions to the chaotic environment input is the life effect. (DNA is an extremely sensitive molecule to the effects of the environmental energy inputs and reacts accordingly)

Differentiated type cells, due to the function they perform and the evolved methods of energy pulse activating inputs have various lifetimes, ranging from a day to a lifetime (Brain cells in mammals) and therefore they have varying susceptibilities to damage as for instance the maintenance cells, supporting the brain cells reacting to mental related input events, become damaged through aberrant energy inputs and this damage is copied when the cells are propagated (Ageing) and they are unable to continue efficiently cleaning up the excess synaptic protein emissions and therefore the brain cells with a mental output are choked and damaged, even destroyed and then dementia, Alzheimers etc. occurs.

Ref

1. The Science of Zoology., second edition, page 60. Paul B. Weisz McGraw-Hill Book Co
2. One Face, One Neuron. Scientific American Oct 2005 by Diane Martindale.
3. Environment Exposure Divergences. Collated by John A. LeRoy
4. Histones. Scientific American

THE DIFFERENTIATION OF GENES (CELLS)

(The "HOLY GRAIL" of Biologists etc)

Reference   http://www.newscientist.com/article/dn3548-electrifying-claims-for-dna-are-dashed.html

Extract from reference

"Quantum tunneling

So how can DNA conduct over short distances, but fail for longer ones? The molecule's short-range conductivity depends on electrons moving between base pairs that run down the centre of the double helix.

To do this, they overcome the associated energy barriers through a mechanical effect known as tunneling. And according to theory, tunneling becomes increasingly unlikely as the distances involved grow longer".

NOTE

The "Differentiation of genes" is the physical reaction of specific genes to specific energy pulse inputs of fixed frequency associated with the species environment. This results in the initiation of the production of damage control outputs with the required outcomes from the species types of various evolved cells ensuring its survival in the environment.

These cells with the genes incorporated in their DNA are initially evolved as a reaction of the DNA to additional damaging effects of the specific energy consistent with persistent environmental change. These energies causing the establishment of genes etc are over and above the basic energy input of heat that initiates the activation of the DNA chromosomes and installation of genes etc during propagation. The resultant adjusted output of the cells, a protective, survival process resulting in the establishment of and adaptation of the species (evolution) and the characteristic of being compatible with this continuing complex energy input supports a coordinated survival effect with the output of the associated cells. The species is then in a situation where its DNA's overall reaction to the complex environmental energy effects results in its survival in the environment.

Almost every somatic cell of a specimen has the same DNA (The only entity in any species that reacts and this is the result of an energy pulse input) and hence the same genes, Regulatory Memory etc. The differentiated outcomes of organ cells etc. occur as the specific genes and their associated Regulatory Memories can react only to the specific energy frequencies responsible for their installation in the first place as they are now tuned to these frequencies. This is a result of the cells having been evolved to accommodate only specific incoming energy processes i.e. direct energy, chemically controlled (From sustenance) or a mixture of both that access the specific genes etc of the dedicated cell with the genes then being activated and reacting.

Being the only entity that reacts to the environment the DNA (A chemical molecule) carries all of the genes etc that are selectively available when required and as a cell is duplicated the DNA is progressively established with the differentiation process not activated until the genes are installed (A heat energy vibration process on the DNA). This overall process (Meiosis) is time consuming and in sequential steps and consequently if a physical intervention of the installation of the required environmental energy effects takes place on the DNA, prior to the natural occurring effects the cell will be differentiated as required. (This is confused with the so-called Stem cell effect, which is an illogical concept and in truth is a non- event.)

The ability of the cells of the human species and any species to produce a variety of outcomes associated with evolved characteristics of organs etc are controlled by this property when activated by the particular environmental energy pulse associated with the establishment of the organ etc, thus allowing survival of the species in this environment. The genes etc are initially installed, activated and differentiated during the conception stage and are updated prior to the next conception process as environment change occurs ensuring survivability of future generations in any persistent change of environment.

HYPOTHESIS

The DNA is the only type of chemical molecule that is present in all species of life that responds to an outside influence (Environmental energy pulses) that initiates all outputs from the cells (These responses may be activated by chemical means as the chemical undergoes a reduction process within the cell, releasing controlled energy pulses to a fixed frequency. The nucleotide pairs of the DNA are always constituted of the same chemical elements and have been established in an order that makes them susceptible to resonate harmonically in response to particular energy pulses emanating from the Environment in a selectively controlled manner. In the presence of a magnetic field they produce an electrical current as they vibrate in response to this environmental energy pulse. This current induces a force field, varying with the intensity of the environmental energy pulse, that activates the necessary response from the associated gene and hence the cell. In the event of a change or addition of a similar new energy pulse (Environment change) an adaptation in response to damage takes place with a change of or addition to the nucleotide pairs of the DNA that are then tuned harmonically in a responsive arrangement to the frequency of this new energy pulse. When the new energy pulse is an addition to the existing pulses this causes the initiation of additional cells with an adapted response and eventually the addition of or adaptation of an organ, and along with the DNA host cells tending to or always making the species compatible with its environment as their outputs counter the damaging effects of these energy pulses. As such the environment change is the cause of evolution and the continuing response of the cells of all descriptions to the chaotic application of energy pulses from the species environment is the "Life" effect with the DNA the "life" entity.

THE SCIENCE OF LIFE AND EVOLUTION

Without the chaotic application of the energy effects of its environment there would no continuing reactions from the DNA to adapt and therefore no "life" effect.

NOTE

When an environmental change results in the disappearance of a particular energy effect the Gene is retained in the DNA, as it is a result of the permanent strain due to the stress resulting from the environmental energy and the replication process of the DNA does not accommodate its elimination. The gene(s) then become unresponsive.

The Para "Quantum Tunnelling" of the Ref reflects the DNA's restricted ability to conduct electricity over short distances. This effect was discovered by Nanotech engineers and they missed the significance of it in relation to the functioning of the DNA, genes etc.

The property of conducting electricity over isolated short distances allows lengths of DNA to function as a series of electrical alternators. These "alternators" are set up by the addition of frequency-tuned nucleotides (The Regulatory Memory, RM) associated with the appropriate genes etc as the DNA extends and evolves to cope with additional environmental energy effects the species is being subjected to.

The persistent changed environmental energy pulses causes a process to result of destroying associated nucleotide pairs now out of tune with a damage control reaction eventuating. This reaction results in additional nucleotide pairs, able to harmonically resonate, being included into the DNA, thus ensuring no further permanent damage effects to the DNA from this ongoing energy pulse. As the destruction to these DNA nucleotides takes place, adjacent nucleotides are also strained beyond their elastic limit (This strain effect is a record of the damaging effects of this particular environmental energy pulse and this length of DNA with strained nucleotide pairs is the "Gene" whilst the section with the harmonically tuned nucleotide pairs is the RM and it reacts with a "switching" signal reflecting the changing influence of the environment).

The new energy pulse of this specification, being still incoming causes these harmonically tuned nucleotide pairs to respond to the specific pulses of fixed frequency causing these selectively tuned nucleotides of the DNA to harmonically oscillate through the magnetic field that is ever present in the cells of all living species. (The energy pulses are provided to the DNA by one of three systems, direct environmental energy impulses, secondly breakdown of chemicals, that are produced by cells as a result of direct environmental energy, activating downstream evolved cells, by releasing controlled energy pulses when chemically reduced within the cell (A conduit system for conveying the effects of the environmental energy), or thirdly a combination of both).

For chemical dependent cells a system of receptors has evolved to receive chemicals relevant to cells evolved functions and on breakdown of these chemicals within the cells, signature energy pulses are released triggering the DNA to respond causing the appropriate output from the cell e.g. pancreatic cells producing insulin. When the relevant cross members (Nucleotide pairs or ladder like rungs, base pairs) of the DNA vibrate through the magnetic field provided by the radioactive carbon and potassium that are always present in the cells of living species electricity is generated by the RM associated with a particular function. When this occurs and the energy pulse is associated with a gene, a fluctuating energy field is generated around the conducting members of the RM, the section of the DNA capable of being electrified associated with the gene, thereby signaling a controlled reaction from the gene, with an enzyme assisted RNA chemical (Ribo Nucleotide Acid, a half copy of the genes DNA) copying process. When this signal identifies the gene with appropriate definitive properties, the RNA is formed and then acting as a messenger, the RNA activates a ribosome action to produce proteins, hormones etc adjusted to allow the survival of the DNA in the "Chaotic" yet persistent damaging effects of the environment. Multiple energy pulses of different frequencies may be present and the resultant vibrations of the various nucleotides reflect the strength of the pulse and therefore generate electric pulses creating surrounding energy fields that are variable and representative of the environmental effects on the specimen and hence its genes and for brain cells the responsive DNA to mental effects. The result is controlled outputs from the cells that are continually adjusted by the responses from the cell's DNA, genes etc to the relevant energy pulses as they vary in strength. The ongoing responses (The life affect) system ensures the survival of the specimen in this environment. This is differentiation of the genes etc and how it occurs.

The restricted length of conductivity prevents the electric current being dissipated over the length of the DNA chromosome and therefore positive signals are delivered to the relevant section of the DNA, genes etc.

For further refer to "Theory of Life and Evolution" by John A. LeRoy (Unpublished)

Justification of the facts used in this hypothesis.

FACT	JUSTIFICATION
1. All species have DNA	Established by genetic and biological researchers.
2. All species are compatible with their environment	Observation of fact.
3. All the effects of any environment are passed on by an energy process	Physics and chemistry phenomena.
4. DNA is a vibrating body responsive to environmental energy pulses	Pjotr Garjajev, Russian researcher experimental result.
5. All cells have a magnetic field	Physics experimental observation.
6. Electricity generated in cells as a result of energy input (stress)	Electrical activity observed in areas of the brain when under a specific stress.
7. Cells are not self-propagating	The property of cells to self propagate is completely contrary to the laws of physics.
8. DNA response is not spontaneous	No physical affect can be spontaneous where there is no initiating influence and if it where so species could immediately adapt to changed environments, which does not occur.

9. Signaling to the genes   Signaling experimentally observed
    eliciting a response from the   drawing a reaction from the genes
    genes etc. due to electricity   etc, believed to be spontaneous
    generation (This Signaling is   by geneticists (Identified as
    compatible with the Energy   "Switching") but recognized as an
    changes of the environment)   electrical engineering phenomena
                                         associated with electricity
                                         generation.

Article from Melbourne Herald Sun

Fatal ailment linked to painkillers

Painkillers containing ibuprofen have been linked to a rare but fatal potassium deficiency.

Ibuprofen is found in many painkillers sold over the counter.

Researchers told of four patients who presented to Australian hospital emergency departments with lethargy, muscle weakness and paralysis.

Each patient was a long-term, regular user of ibuprofen, a drug for chronic pain (popular brands include Nurofen and Advil)

They were diagnosed with hypokalaemia, a dangerously low level of potassium in the blood.

The research is published in today's Medical Journal of Australia.

NOTE

The Potassium contains a significant proportion of K40, its radioactive isotope that provides, along with radioactive carbon (C14), a large proportion of the magnetic field present in every living cell. Reduction of the presence of this element in the cell reduces the isotope present and hence the magnetic field is reduced causing a drop off in efficiency of the cells electricity producing capacity and hence leading to mal-functioning of the cells.

LIFE, ELECTRICITY AND ITS GENERATION

The Fuelling and Igniting (Activation) of Cells

Russian Professor Pjotr Garjajev and his team of researchers, as explained in the paper "Wave Genetics: On the Wave Structure of DNA and Resonant Interactions of Genes and Environment" (http://www.spaceandmotion.com/evolution-biology-wave-genetics.html) by Geoff Haselhurst established that the DNA molecules were susceptible to pulsing energy vibrations emanating from the environment and as a result were prone to resonating.

The fact that DNA molecules resonate when subjected to energy pulses emanating from the environment is confirmed. The physical phenomenon of "Vibration and Resonance" is the most common in the universe. This fact established by them as a result of experiment and observation is the points of interest along with the conclusions

The statement however "So human languages did not appear coincidently but are a reflection of our inherent DNA" is not agreed with and in fact is illogical.

Having demonstrated in the paper "Theory of Life and Evolution" that the DNA molecule has been evolved and developed as a reaction to its material exposure to the incoming damaging pulsing energy effects of its environment it is not logical and does not make sense that the human DNA molecule was already established, as indicated in the reference paper, so that the language was developed as a result of the pre-existing arrangement of its "Junk" nucleotides. This situation also applies to the other energy inputs related to the senses (as supported by the variation of the DNA molecules of various species and varieties of species having different numbers of genes, nucleotide pairs etc.).

As the life process first developed in the early history of the world and under the right conditions, molecules of RNA (half sides of the DNA molecule) naturally formed, due to a chemical process aided by an available enzyme, with nucleotide extensions that consisted of four different types, namely A, C, G and T as they have been designated. Each nucleotide extension of the type A had an affinity to join with type T, whilst the nucleotide extensions C were prone to joining with nucleotide extensions G. Mathematically this meant the chances of the RNA molecules originally forming DNA molecules was restricted, indicating a limited number of pairs originally made up the DNA molecule. The only conclusion then available is this has resulted in the situation where over the billions of years of the ongoing development of the life process the DNA molecules have been extended and adapted by a controlled process when the variations of the environmental energy pulses the evolving species has been exposed to, result in increasing number of genes and variations in their makeup, together with variations of and increasing numbers of "Junk DNA" nucleotide pairs, in the DNA molecules of mobile species, that were and are developing depending on their ongoing exposure to and ancestral history involving the rapidly changing situations requiring physical actions and assessment with resulting inherited mental reactions and traits leading to survival in these events.

NOTE

Once induced into the DNA molecule a gene is always present even though it may be dormant due to the environmental related energy pulses originally involved having disappeared as the environment of the species has undergone change. A gene is installed by exposure to environmental energy impacts that damage the nucleotide pairs beyond their elastic limit and then the gene reflects the damaging capability of this aspect of the environment. During the installation process of the gene its nucleotide pairs have become coded resulting in the ammonic acid molecules of the protein it initiates being produced supporting survival of the DNA molecule and hence the species in the environmental energy pulses. When the cell undergoes propagation the damage effects that are the genes in the DNA molecules are perpetuated by a further damaging and copying process as they are reinstalled by a physical process of energy pulses that the renewed chemical molecules are subjected to.

As an evolving species we are receivers of environmental influences (always energy pulses) that are the initiators of the life process via the DNA molecule that is extended and revised including the addition and extension of genes by a damaging hit and miss energy driven replacement and addition process resulting in reactions from the evolved cell and consequently of the species as the environment variation results in development of characteristics, that lead to protection, healing and survival of the species via its DNA molecules in these potentially damaging conditions.

Where Eukaryotic (Multi cell) species are involved the energy pulses initiating reactions from the cells may consist of direct energy inputs, such as those associated with the senses or internal cells by carbon based chemical molecules with a range of energy bonding's of the atoms that on controlled chemical reduction within the cell are released as energy pulses of fixed frequency e.g. sugar when reduced within the cell releases approx. 140 pulses per molecule. Further some cells may be subjected to a combination of both sources.

The sugar molecule is utilised to provide the basic energy requirements in all cells and does not play a part in other specific outputs other than those cells of the pancreatic organ where it is influential in the production of Insulin.

Where further carbon based molecules are reduced within a cell they are relevant to the differentiated type cell and on chemical reduction within that cell release energy pulses of a frequency that activates the responsive gene(s) of this cell and this results in the initiation of an appropriate output from that cell

When the carbon based chemical molecules are involved they are conduits for the environment effects, having been grown and nurtured as a result of the incoming energy pulses from the environment

When continuing similar persistent incoming energy effects are present the result of is the incoming vibrating environmental energy of fixed frequencies identifies and resonates component parts of the

DNA molecules associated with the gene (RM) initiating a continuing similar physical response and similarly a mental reaction output from the "Junk" DNA as a result of the effects of the incoming energy pulse impacts from the senses on the glia cells becoming recognisable due to the previous strain on the coded "Junk" nucleotide pairs having lingering, time sensitive distortions due to the material recovery rates making short term, medium term and long term memory possible as the incoming energy pulses access them, and send signals to the relevant neurological cells, where upon action signals are initiated by electrical pulses via nerve fibres to the muscles etc.

There are supporting genes evolved within the glia brain cells to produce protein that forms pathway through the synapses for conducting the electric generated pulses and support cells dedicated to cleaning up the protein responsible for protein Tangles etc. When these cells loose efficiency the clogging effect of this protein causes death of the glia cells with the onset of Dementia, Alzheimers etc.

As the language developed it was due to the range of environmental noise energies representing danger or survival increasing when the species evolved to hunt on the ground via two limbs in the upright stance, so the "Junk" section of the DNA molecule relating to noise increased by the hit and miss process as did the genes relating to the supporting larynx etc. evolve.

The junk DNA is the portion of the molecule that has been developed without any extensive arrangement of genes, as incoming energy pulses associated with the mental processes such as noise energy is not usually persistent and extensive enough to permanently damage the nucleotide pairs similar to those being vibrated as a result of incoming energy pulse effects from physical aspects of the environment and causing genes, through the extra damaging effect, to evolve.

As it is not necessary that an extensive physical output by the genes be available to counter the damaging effects of the noise energy for survival, it has resulted in the DNA molecule evolving to utilize the sound energy by recognition (tuned to) of preinstalled, but recovering distortions (A material property) of nucleotide pairs (Forming the "junk" section of the DNA molecule) produced by the effects of similar previous sound energy pulses and with vibration occurring to these nucleotide pairs energy pulses that are produced are passed on via the synapsis' of the glia (brain) cells with the passageway enhanced by temporary protein injected into them as a result of the activation of relevant previously evolved support genes. This process provides a pathway in a progressively condensing fashion to the designated pre committed neuronal cells where they are integrated with the relevant energy pulses of the other senses to guide and alert the species to the presence of events, originally sustenance and danger, where reactions lead to physical survival of the species and consequently the DNA molecules in the environment.

Speech (i.e. language) is a characteristic developed when sound energy inputs resulted in the associated pulses evolving the Regulatory Memories (RM) of the DNA molecules (In this case the evolved responsive "Junk" nucleotide pairs) enabling cells with a response capacity becoming coordinated with the other sensory inputs and the conscious memory search and thinking processes leading to survival (We humans think in our language). The more complex the specie' developing environment becomes, the more the species ability to cope and react is evolved.

The harmonic arrangement of the "Junk "Nucleotide pairs observed in the reference paper is that evolved for a response to the frequency of the particular incoming energy pulses associated with the senses.

The presence and flow of electricity in a species makeup has been observed and established by experiment.

Electricity, with the capacity to transfer energy does not just appear but must be originally generated and this involves the initial application of energy from a source, the presence of a magnetic field and the capacity for an electricity conductor to be repeatedly moved systematically through this field whilst being connected into an electricity conducting circuit.

With a magnetic field being available in each individual cell as provided by the radioactive isotopes of Carbon (C14) and Potassium (K40) that are always present, the remaining attribute of a conductor moving through this field systematically is provided by the incoming pulsing energy impacts accessing the relevant DNA nucleotide pairs

Due to the pulsing environmental Energy impacts of fixed frequencies impacting the now tuned nucleotide pairs (Random Memory (RM)) associated with the genes of the DNA molecules generate a confined electrical current that switches on the initiation by the gene of the cells physical output by way of the force field generated that surrounds the nucleotide pair conductors as the current generated in them rises and falls. This controlled electrical signalling production has been enabled by the restrictions of Quantum tunnelling characteristic of the DNA carbon based material assisted by the fact they are in the presence of water

With the incoming energy pulses from the environment accessing the senses the relevant coded arrangement of the "Junk" nucleotide pairs of the brain cells DNA are now resonating through the magnetic field generating an electrical current, however it becomes apparent they now have evolved to download the electrical pulses generated via the synapses to the neurological cells thereby gaining a reaction.

NOTE

Quantum tunnelling is the characteristic of a conductor where it conducts electricity only when in a state of strain and this characteristic increases as the strain increases.

For the normal differentiated non-brain cells the electricity flow is therefore constrained to these conductors and there is a force field generated, surrounding the nucleotide conductors that rises and falls commensurate with the incoming energy pulses and these pulses are commensurate with the environment when delivered directly and in the case of secondary delivery (Chemical) where they are reduced within the cell releasing energy pulses these chemical molecules are conduits, under normal circumstances, for environmental effects.

The effect of this force field is constrained to the relevant gene. The gene nucleotide pairs as they are in a different physical state (Strained) to the normal DNA nucleotide pairs of the molecule are sensitive to this force field that rises and falls reflecting the environment and it "switches" them on, producing a half sided copy of the damaged nucleotides of the Gene (mRNA) that then initiates the cells output. Due to the hit and miss process involving damaging effects of the incoming energy pulse impacts and the enzyme assisted chemical renewal of the nucleotide pairs until the coding of the gene is such it has reached a state where it initiates the production by the cell in its cytoplasm of a protein, hormone etc. output that then counters the damaging effects of the incoming environmental energy pulses responsible for the damage to the gene and provides survival counter effects and this leads to the species survival in the environment.

An example of this is the protein, chromatin that is wrapped around the DNA molecules, with the characteristic of muffling the incoming energy pulse impacts relevant to the particular differentiated cell type. This muffling effect is developed to a point where the incoming impact is controlled to the degree that the cell can function efficiently without being significantly damaged. (This illustrates the Physics law of Cause and Effect).

In the case of brain cells, Glia cells are utilised where the capacity to download the recognisable effects of the generated electrical current has evolved, to be downloaded and passed to the relevant neurological cells where an appropriate response is initiated.

An incidental fact from this is that illnesses arising from a potassium deficiency, sometimes induced by the drug Ibuprofen, are related to the fact the magnetic field required for the efficient operation of the cells is deficient as the potassium is eliminated.

06 PROTEIN AND ITS ROLE IN DISEASE

◇◇

Proteins are the building blocks of life, where the DNA, a complex chemical molecule is the major player in the process of life. The DNA reacts to incoming environmental energy pulses it is exposed to that result in the initiation of an output of protein from its cell providing a system of control against the damaging effect of these energy pulses on it, the DNA, leading to its survival. As the various incoming energy pulses of the environment are normally delivered in a chaotic manner its continuing reactions in response to keep up with and counter the potential damage are the apparent life effect with consciousness being originally due to the reaction characteristic where mobility was necessary for the acquisition of sustenance for energy supply to the cells of the evolving eukaryotic cells and hence its DNA and therefore the species to survive in this environment.

Where a persistent change of environment occurs this entails an energy change affecting the DNA. Damage results with the DNA being activated to stabilize in this environment and as an aspect of the controlling measures protein is produced to protect against this energy effect and as a consequence the formulation of the protein is synonymous with the characteristics of the involved energy. An individual of a species having ancestors who have experienced slightly divergent exposures to persistent energy effects inherits resultant slight variations of the genes nucleotide pairs that stipulate the formulation of the various proteins etc it produces. The complex molecular makeup of the proteins has virtually no limit to the arrangement of its atoms and sub molecules and this is controlled by the individuals nucleotide pairs arrangement of the individual genes (A gene may consist of 3000 pairs of nucleotides, evolved over thousands of generations due to varying exposures to persistent environmental changes) and as a result each individual of a species has its own formulation of proteins capable of sustaining its survival, however where the environmental exposure has been similar the proteins are generally similar and the individuals can breed without difficulty. Where energy inputs of significant divergence have occurred effecting the earlier generations of branches of a species over long term due to the variance of their habitats and environments they have been exposed to, then the protein produced by individuals of the two populations will be considerably different. When a couple from these divergent populations produce an offspring then this individual may inherit genes involved in the production of proteins not acceptable to the immune system, which has been inherited from the other parent that evolved to cope with problems associated with this parents basic environment. The situation is tantamount to the rejection of a transplanted organ with foreign protein, where the proteins evolved in response to a variation of environmental energy and therefore having genes producing proteins that are not compatible with the recipient's immune system

In the beginning of the life phenomena small molecules of DNA formed naturally and with their format of a ladder like structure and dimensions being such, some of the molecules of the DNA were compatible with the frequency and amplitude of the energy pulses of the environment i.e. heat and light, that when applied caused vibration and with their rungs (nucleotide pairs) being distributed in a manner that were harmonically responsive to this wave energy the DNA molecule survived the damaging effects of this energy. (The remaining DNA molecules were destroyed by the pulsating energy).

With the passing of time a miniscular addition to the environmental energy the DNA was exposed to occurred, damaging several of the nucleotide pairs (The Rungs), as they were no longer in tune with these energy pulses and degrading them. This resulted in a chemical repair process developing that eventually established nucleotide pairs that were in tune with the energy pulses. This occurred as the break down of the nucleotides facilitated an enzyme assisted energy / chemical process to occur, combining the now available elements with natural occurring background materials

As this process developed a scab like formation (Protein) protecting the remaining DNA from further damage, formed and the makeup of this protein chemical was established in response to the frequency and amplitude of the energy pulses initiating its production. The arrangement of the atomic elements of this protein molecule formed to be synchronized with these energy pulses i.e. where the energy pulses (Wave energy) could pass through it without damaging it but where the energy level was degraded. Part of the protein formed is the chromatin (Epigenetic) coating and the Telomere end capping of the chromosomes (Applicable to Eukaryotic species)

The initial Prokaryotic cells with their single strands of DNA were mainly exposed to the direct environment for their source of energy and with their intermittent exposure to this energy they multiplied and evolved slowly as the damaging pulsing effect on the DNA eventually caused material fatigue to the hydrogen nucleotide bonds causing it to split in halves and a chemical reaction initiated to match up the nucleotide acids was set in action with the DNA being reestablished when rebuilt into two separate cells, and further establishing the process of protein, and replacement nucleotide acid production enabling its survival in this changed environment

In the early history of the advent of Eukaryotic species with pairs of chromosomes eventually developing Prokaryotic single cells were infected by mitochondria energy producing cells and eventually further similar prokaryotic cells, resulting in a process of energy extraction on a continuing basis from sustenance (Ref 2.) with the additional energy effect significantly increasing vibration of the DNA within the cell causing the split of the nucleotide hydrogen bonds at a rapidly accelerated rate and therefore rapidly increasing cell production. This resulted in the prolific generation of protein, hormones, enzymes, lipids, sugars etc as the DNA nucleotide acids reactions established evolving sophisticated characteristics at a more rapid rate. This process continued and continues over generations (Evolution) with cells combining and multiplying as a coordinated reaction to the effects of the incoming variations of the environmental energy until numbers of divergent purpose cells (Differentiated) existed to match the demise rate of like purpose cells resulting in maturity of the organ etc these cells were involved with. As the specific incoming environmental energy stresses were now becoming more complex and changed they were delivered by way of direct energy effects as well as sustenance (Foodstuffs, raised and evolved in the same habitat, with protein, sugars and lipids etc contents now conduits for the equivalent energy effects when chemically reduced in the specimens cells) resulting in cells developing with evolved DNA and a variation of output reactions, as an extension of the existing cells due to these slight divergences. This established differentiation of the cells with the DNA of the cells being committed to react to specific frequencies of various energies, and developing membrane receptors to receive the involved chemicals with their energy pulse releasing characteristics when chemically reduced. This situation led to the bulking up of and evolution of various species. Over the period of evolution, the members of generations of an evolving group were exposed to generally similar environmental conditions, however individuals of the group underwent slightly varying exposures and this eventually resulted in the members of a species having generally similar proteins, but all individuals having unique proteins commensurate with their DNA's inherited and ongoing reactions to the wave energy of the environment previously experienced and now involved.

The process of an ongoing protein etc development continues due to the exposure of the cell and its specific DNA section (the gene and RM) to the specific environmental energy responsible for establishment of this gene. The production of these chemicals is enabled by the element potassium with a proportional amount of its radioactive isotope K40 and an isotope of carbon C14 being present, (This K is present in every cell and no cell exists without it), causing a magnetic field to be present and when the relevant harmonically tuned nucleotide pairs of an RM associated with the gene(s) of the DNA chromosome(s) is subjected to this additional specific incoming pulsating energy it results in these nucleotide pairs harmonically vibrating (The Regulatory Memory (RM), is produced initially in response to this energy frequency) causing an electrical current to be generated in them. A fluctuating energy field surrounding these active "conductors" then represents the varied and varying environmental energy effect and thus the adjusted environment the relevant DNA length is being subjected to and this energy field highlights (Switches on) nucleotide pairs adjacent to the RM nucleotides that have also been subjected to damage, by lose of their resilience, but not destroyed. These damaged nucleotide pairs are thus non-responsive to the incoming environmental energy pulses, but represent the damaging effect of this slight deviation of the energy effect that is then countered. (These damaged nucleotide pairs are the genes). This switching effect, synonymous with the strength, amplitude and frequency of these energy pulses effecting the nucleotide pairs (RM) results in the identification of the relevant gene nucleotides further resulting in the initiation of an ongoing chemical production of varying quantities of a specific protein (along with other chemicals) guided by the genes makeup, and this unique protein is synonymous with the environmental energy responsible for it and in its many forms serves a protective function and the building blocks of the species. (The possible atomic arrangements of the molecules of proteins have enormous diversity and as such are established to control and dampen the passage through them of the specific wave energy pulses responsible for their establishment)

As the species matured the number of cells duplicating was due to DNA chromosomes suffering fatigue at the hydrogen joints of nucleotide pairs due to the vibrating input of the environmental energy pulses it was being subjected to and splitting with an enzyme aided chemical reaction replacing the missing halves of the DNA chromosomes resulting in the cells duplicating. As the number of cells increased some were dying off until the rate of renewal equalized with the demise rate. This is maturity and as the quality of the renewing cells dropped off and the demise rate exceeded the renewal rate this becomes the ageing phenomenon. The above is a simplified explanation relating to a particular environment change, however many aspects of the environment may vary as it becomes more complicated with additions and deletions to the environment energy affecting the species and similar processes related to these environmental changes occur over long periods of time and this results in the cohesive evolution of the species characteristics as their cells DNA is manipulated and adjusted leading to survival.

Evolution then is the adaptation of the species DNA and hence its characteristics (features) as they are subjected to change by the persistent environmental effects it is exposed to resulting in its survival as it reacts, providing the resistance to withstands the damaging effects. This encompasses the use of the protein etc in many forms and functions to provide survival features. With the passing of time the evolved characteristics of humans have become more and more complex as their contrived environments becomes more sophisticated, with the process always tending to achieve survival. The contrived environments of humans however are not always in sync with the basics of the process, leading to problems.

With additional changes to the environment additional RM's and genes were added and the system was repeated, with a differentiation system occurring of dedicated cell membranes being controlled by the different energy inputs responsible for their development and producing an output from dedicated arrangements of cells (Organs etc) leading to the survival of the DNA and therefore developing species. This integrated system of the DNA continually expanding and responding to the potentially damaging effects of the continuing chaotic energy effects of its environment, when its characteristics (or features) are generally pre committed by the extensive ongoing exposure and adaptation of its DNA in previous generations is the phenomena of life and evolution. (Ref. 2.) This controlled response to the incoming energies of a species environment is the "Plan" built into our DNA. Mobile creatures, in particular humans, are aware of the phenomenon of life, as the DNA has been evolved to provide direction in a conscious and logical manner to achieve survival. The degree of awareness is related to the number of specific brain cells that have been evolved enabling specific reactions to the incoming information via senses, and the quantity of evolved dedicated "Junk" DNA in each cell.

As species are exposed for long periods to an environment their DNA genes etc are evolved (extended and /or genes added) as a result of the applied frequencies etc of the energy of any persistent environment changes they are subjected to. This change of environment also leads to their divergence when they gradually migrate from their original habitat with sufficient time elapsing to convert their DNA; genes etc (evolve) to cope with these conditions as they produce adapted proteins.

This accounts for the variations of proteins that are unique in molecular formulations present in each individual of a species. (Ref. 1) The protein etc is generally similar from a species comparable organs etc, but with variations in detail depending on the individuals and its forbearer's history of environmental exposure. The greater the historical divergence of persistent environmental exposure for mating specimens of the species the greater the likely hood of the output of genes etc being incompatible in their young leading to such diseases as MS,

Cerebral Palsy, Diabetes 1 etc as their immune system attempts to correct the perceived problem. (The inherited Immune system, having been evolved due to the specific environment exposure of one parent may be incompatible with the output of specific cells whose active genes have been inherited from the other parent) The problem does in fact range across all characteristics, both physical and mental of all Eukaryotic species, but may vary in members of a family in accordance with Mendel's Laws of Inheritance as they randomly inherit reacting genes etc from either parent.

With the evolution of brain cells and their role of recording and downloading information to the particular action neuronal cells with their synaptic connections with associated proteins being produced by the involved cell being injected into the synaptic gap to conduct only the relevant energy pulses associated with this information. As the input of specific memory information becomes more regular the protein input becomes more consolidated in the synapses eventually resulting in a funneling effect, establishing a pathway with instant recall and therefore rapid reaction. The protein produced by each cell is specific to the energy effects applied to its DNA and therefore it conducts the specific energy only involved in its installation. Where the incoming energy effects are not persistent the protein is dissipated and recycled by supporting cells. When ageing occurs and these supporting cells loose efficiency this excess protein causes clogging of the brain cells and the demise of many of them. This is Alzheimers, Dementia etc.

The situation of specific energy inputs accounting for the development of specific genes and hence proteins also accounts for the rejection encountered in limb and organ transplants.

NOTE

On first reading this paper may give the impression that it is conjecture, however the facts and physical phenomena brought together are at least ninety percent or more established by scientific experimental observations, but simply not brought together in the literature because their implications have not been recognized.

Reference

1. The Science of Zoology (Second edition) by Paul B Weisz. (Biology Manual McGraw-Hill book Co)
2. "Theory of Life and Evolution" by John A. LeRoy (Unpublished)
3. "Making Memories Stick". R. Douglas Fields. Scientific American. Feb 2005 (Copy Attached)
4. "The Hidden Brain". R. Douglas Fields. Scientific American Mind. May/June 2011 (Copy Attached)

P.S. Chapters 4-6 of Ref. 1. Describe much of the technicalities of the process; however page 113 of chapter 6. Ends with the sentence, "Thereafter, some-so far unknown-stimulus brings about change in chromosome activity". What this stimulus is is established in this paper and at Ref.2. This recognized stimuli (environmental energy pulses) when associated with known physical phenomena, brings logical conclusions to these chapters.

EPIGENETICS, and the involvement of Chromatin Protein in EVOLUTION

◇◇◇

This paper examines the science supporting the hypothesis underlying the Chromatin protein surrounding the DNA chromosomes in a cell's nucleus is a reactive entity (Epigenetic) in parallel to that of the DNA molecules and concludes there is no justification whatsoever, to reach this finding, but concludes that in the absence of the knowledge of how a cell functions perceptions have been substituted to replace facts. Further it is concluded the necessity for epigenetic reactions to initiate outputs from cells to support survival of the specimen is impossible and as such life under these circumstances does not exist.

At the beginning of the life phenomenon, some four billion years ago DNA molecules appeared when two naturally forming chemical molecules (RNA) with nucleotide acids A, C, G & T as part of their makeup formed. These acids, had protrusions with their ends having an affinity for hydrogen atoms, whilst at each side of the acid molecules an affinity to bond with a sugar molecule existed. Eventually these affinities to bonding led to the protrusions acting as jointing points resulting in a DNA molecule with a ladder like format with jointed rung effects forming from the compatible protrusion extensions of the molecules A to T and G to C whilst the bodies of the acid molecules bonded to the sugar molecules forming "rails" (See paper "Theory of Life and Evolution"**). Initially the DNA molecules were in an inert state, however when an arrangement of the DNA of several nucleotide acids AT and GC eventually occurred that was harmonically compatible with the incoming environmental energy pulses it was exposed to (Heat) they began to resonate and the DNA molecule began to writhe as the nucleotide pairs vibrated due to the effect of the environmental energy pulses on it (Vibration and Resonance phenomenon, Ref*). The incoming heat energy pulses continued and eventually a variation or change of the base environment occurred and the initial nucleotide pairs did not have a suitably arranged format to respond harmonically to these energy pulses, leading to damage and destruction of nucleotide acid pairs at vulnerable points of the DNA molecule resulting in chemical residue in the form of an enzyme that had the property of accelerating the production of replacement acid nucleotide pairs from the naturally occurring sugar (Established as being present as it formed part of the original RNA molecules) and the remnants of the destroyed nucleotides and available elements. Additionally a chemical process was initiated producing a complex chemical molecule (Protein) that surrounded the molecule of DNA. (Scab like) that provided protection for the DNA molecule from the damaging environmental energy pulses. These protein molecules consisted of Amino Acid molecules joined together by peptide bonds. The arrangement of the Amino Acid molecules were and are established in an helix spiral within the protein molecule and have the property of muffling the incoming energy pulses of a particular frequency, (An example of one of many is this phenomenon occurring as infra-red energy pulses pass through a glass pane whilst ultra-violet energy pulses do not, even although the ultra violet energy pulses is regarded as the stronger of the two.) whereas the energy pulses of a variety of frequencies may pass through it to a much greater degree.

Protein molecules are not stable over a period of time and as they are continually deteriorating the energy pulse involved with the establishment of the genes activating the production of the chromatin protein continually penetrates through the protein to the DNA nucleotides in a muffled condition reactivating the relevant genes (Ref ***) and re-establishing the protein molecules (Chromatin). This then is close to a stabilized condition and the genes utilized in activating the cells functional outputs activate the production from the cell in a similar controlled manner.

During the formation of the gene (Ref **) as the incoming energy pulses change (Persistent slight change to environment) the nucleotide pairs were and are adjusted by a hit and miss process resulting in a variation of the arrangement of the chromatin protein molecule, and this persists until the "coding" of the DNA molecule produces a protein such that the effects of the incoming relevant energy pulses are controlled. (Three pairs of nucleotides are recognised as forming an instruction (Alphabet) in the formulation of the protein).

Where a variety of several genes have been evolved (Ref**) that are responsive in the cell the chromatin consists of a variety of protein molecules, providing controlled protection for the DNA chromosomes. Eventually over time the vibrating effect of the incoming muffled energy pulses eventually cause the nucleotide pairs to split due to material fatigue. An enzyme assisted chemical reaction now results in a replacement process for the detached half sides of the original DNA molecule. These replacement molecules (RNA) match up with the exposed ends of the nucleotide protrusions, reattaching through the medium of hydrogen atoms. Incoming energy pulses then reinstall the genes to the DNA molecules (Ref**) with chemical processes rebuilding the cell membranes, cytoplasm etc. (Cell propagation)

As the cell is now in the pluripotent stage (Most cells have a common range of DNA molecules) and as it is associated with cells of specific outputs it is adapted by the available specific energy inputs via carbon based chemicals associated with these cells leading to it becoming a differentiated cell of the prevailing type.

The establishment of the differentiated cell occurs, when they are exposed to the chemicals and/or energy pulses relevant to the functioning of this type cell. Receptors are induced in the cells membrane, suitable to accommodate these chemicals allowing the ingress of the specific chemicals to the cells cytoplasm, where they are chemically reduced. (The chemicals are (or should be) conduits for the environmental energy relevant to the specimens environment and for the evolution of this differentiated cell). This results in, on ingress of the chemicals into the cells cytoplasm, energy pulses being released commensurate with these particular chemicals. This energy not only provides the energy required for the cells material output but the released energy pulses activate the specific genes (Ref**) involved with the output of this differentiated cell type. The installation of the receptors is time consuming. (At this stage man induced interference can take advantage of the time delay and unknowingly expose the cell in its pluripotent stage to the energy pulses associated with another type differentiated cell. It is then concluded, quite illogically, that stem cells exist).

The relevant incoming energy pulses activate specific genes via the histones, which protrude above the chromatin protein level, enabling the reaction of the DNA molecule to the energy input to initiate production of the protein, which provides the necessary protection from the energy pulses, ensuring the stability of the DNA chromosomes and hence the cell.

The protein molecules then have the individualistic arrangements of the Amino Acid molecules relevant to the specimen's individual particular evolved differentiated cell and they reflect the specimen's historical heritage.

As there are approx. 220 different types of cells in a human's makeup, there is the same number of a variety of chromatin protein molecule mixes. On observation of these variations it is not scientifically justified to conclude that they have a role as an entity superior to that of the DNA and its genes to play in the function of the cell. They are proteins produced by the cell in a response to an initiating reaction of the DNA genes to the environmental energy pulses, (Primary or secondary) and as every protein, enzyme etc. is, they are produced to play a role in the survival of the specimen in its environment.

The Epigenetic function assumed to be present and to be involved in activating cell function and to cause illnesses and diseases when discrepancies of the protein molecule in a single cell, in comparison to a similar cell, are observed is due to problems with the gene(s) of the DNA molecule that is activating its production. The discrepancies are caused by the response of the cell to the pre-disturbed coding activation of the DNA genes by the environmental energy pulses and resulting production of increasingly faulty chromatin protein allowing increasing disruption to the DNA nucleotides due to the now increased access for the energy pulses. It is a cyclical effect where the damage may slowly increase over generations of the cells as the increasingly damaged nucleotides are copied.

Further to this the protein is not reactive to an outside influence and not present when the cell's DNA first becomes responsive (The cells DNA actually initiates the production of the chromatin protein doesn't it?) and therefore is not an entity driving the life process but is a product produced at the initiation of the evolved capacity of the DNA molecule to ensure survival. The fact is the usual concept envisaged of Epigenetics, without the dampening effect provided by the chromatin protein; the DNA would be so badly damaged the life process would not be possible. When the nucleotides of the gene(S) are damaged by man induced problems, such as aberrant radiations (Energy pulses not normally present in the environment and therefore able to penetrate the protein evolved with specific properties) and chemicals such as asbestos, nicotine etc. with the DNA nucleotide pairs coding being disrupted an unsuitable protein is then produced by the cell allowing the ingress of damaging energy pulses may result in the cell and its propagated cells eventually going into an uncontrolled energy consuming spiral, with developing dysfunctional cells that are running amok as they tend to become parasitical organs (cancer).

Conclusion. Where the environmental energy change is small and persistent and generally occurring to numerous members of a species, changed species characteristics may occur as a result of the environmentally induced genetic changes to the DNA molecule with inheritance factors becoming involved with evolution occurring (Ref *). However if the identified changes to the chromatin (Epigenetic Chromatin) are attributed to causing inherited disease there is no viable identifiable reason, whereby a produced protein can react causing the problem. (Once again, as with the dubious and totally impossible dogma that cells react spontaneously, it is being assumed the protein reacts spontaneously, without the slightest scientific justification or common sense logic being applied).

The identified variations of the chromatin molecules of a differentiated cell type are not the cause of problems but are the identifiable signs that there is a problem with the DNA molecules of that cell.

References

* Pjotr Garjajev, Russian biologist and team, see http://www.spaceand motion.com/evolution-biology-wave-genetics.htm
** Theory of Life and Evolution
*** http://www.newscientist.com/article/dn3548-electrifying-claims-fordna-are- dashed.html

THE FALLACY OF STEM CELLS

<<><><><><><><><><><><><><><><><><><><><><><><><><><><><><><><><><><><><><><><><>>

All Eukaryotic species have differentiated cells and these cells develop progressively as a reaction of the original DNA molecules of the germ (Zygote) cell when specific environmental energy pulsations are applied (frequencies and strength) (All aspects of an environment accesses the DNA in the form of energy pulsations either by primary or secondary means). The differentiated cells develop in the order of the species evolution as specific environmental energy effects arising from the environment access the DNA, initiating cells with the appropriate characteristics. These energy effects access the developing cells when they are in a state of pluripotency and they activate specific genes etc of the DNA where the RM nucleotide pairs of the gene is tuned to this energy frequency. These genes have been evolved over generations and are present in the common DNA molecules of all somatic cells and are a result of and an indication of the damage caused in the reaction of the DNA to these specific environmental energies as they occurred over preceding generations. When these energy effects of the environment continue, ongoing reactions from the cells initiated as a response by the genes etc continue, resulting with an output from the cells that leads to control of possible further damage effects on the DNA arising from this pulsating energy and hence the DNA survival.

Once established in the DNA the genes remain as a constant factor and where the causal environmental energy input is still present the output of the cell continues unaffected throughout the following generations. Where the offspring have genes passed on in the reproduction process a gene from one parent may be dominant or if a similar gene governing a characteristic is necessary to be present in both parents and become paired during conception before the characteristic appears in the offspring this is "Inheritance" and is involved in the process of evolution. Where the environment undergoes a persistent change or addition the DNA molecules react with the genes etc being adjusted or added to, adapting the molecule and hence cells in survival mode with an adapted output. This eventually results in change or potential development of a specimen and when members of a species are exposed to a similar persistent environmental change they are all subject to change in a similar manner and on reproduction the offspring may be evolved to suit the environmental conditions or carry the genes etc with the potential to do so.

Over the generations of a species evolution, genes etc, that represent the reactions of the DNA to the damaging effects of the environmental energy pulses are established to cope with the many and varied energy inputs. These initiating energy pulses, when a persistent change occurs to the environment now have a variation of fixed frequencies and amplitudes that cause a change to the DNA nucleotide pairs of the relevant Regulatory Memory (RM) and gene and a reaction occurs from the gene that initiates an adapted output of the associated cells. The outputs are in the form of proteins etc of many and various types and they have a chemical format with an atomic arrangement dictated by the gene and hence the responsible energy input.

The resultant output of the cell then protects the DNA molecule against rapid damage from this particular energy or ensures survival in the particular combination of energy pulses from the environment until the DNA nucleotide pairs of the cell eventually reach a fatigue state and fracture at the hydrogen joint. On rupture a state of cell duplication can usually occur and hence ongoing survival. The protein etc along with the diverse cells with their many types of outputs are the species and the more complex a species environment the more complex the species. As any environment may be represented by many and varied energy pulses, more genes etc may be established and as a result more cells with varied functions and outputs are added with the characteristic of coordinating as required with previously evolved cells and this is established through the medium of the proteins etc, produced with the characteristics of supporting the DNA's survival

Cells cannot be and are not self-propagating; they do not and cannot act spontaneously, nor can the DNA split apart of its own volition and reformat itself into the DNA required for the duplication of cells. These three supposed events under any circumstance are totally against the laws of physics and incidentally common sense.

The development and functioning of a cell (differentiated) with a specific purpose is a five-step process beginning with.

1. The accumulation of the various elements necessary for the production of the duplicated cell and of the makeup DNA. Chemical reactions aided by enzyme actions and driven by an energy input ensures the component parts in their relevant chemical formats including enzymes etc are available.

2. During cell propagation, when the DNA of the cell splits apart it is due to the nucleotide rungs splitting at the hydrogen joints as a result of fatigue they have been exposed to due to the vibrating effect of the environmental energy pulses they have endured over the cells lifetime. They have been subjected to a critical number of energy frequency cycles particularly at stress raisers i.e. the RM's of genes associated with the role of this cell.

3. An enzyme-aided chemical reaction initiated by the energy input results in the replacement of the missing nucleotide acid halves of the DNA molecules, with the relevant structure of the cell being chemically reformatted as it divides into two cells.

4. As the DNA chromosome molecules of the cell are completed they are subjected to the vibrating influence of the heat energy input (Heat energy exists at any temperature above minus 273degrees centigrade). As the initial half of the DNA chromosome now carries the previously stressed nucleotides (Half genes) and harmonically responsive arrangements of the DNA nucleotides to incoming environmental energy pulses (Ref 1.), the accentuated vibrating effect results in the genes being once again fully installed, whilst the arrangements of the nucleotide pairs now carry the ability to react to the appropriate environmental energy input (delivered by primary or secondary chemical means). At this stage the cells are pluripotent.

5. At the embryo stage, when sufficient numbers of pluripotent cells are present, stored chemicals (proteins etc) representing specific environmental energy effects (Included in the yolk of the maternal parent egg and updated over her lifetime, whereas the males DNA of the fertilizing sperm has been updated (evolved)

over the paternal parents lifetime) are randomly allocated and the incoming heat energy effect sets the process in motion as it activates the DNA and reduces the chemicals, releasing energy impulses commensurate with specific aspects of the evolving and evolved environment and thus setting the genes etc in action as they react to the differentiating effects of these energy impulses thus ensuring survival of the DNA molecules with specific reactions and thus the species in the damaging effects of the environment. As the species matures, replacement cells / developing cells go through a similar process, but are exposed only to the chemical and primary energy effects now influencing the organ etc they are involved with. At this stage, the relevant chemicals, proteins sugars etc are admitted to the cytoplasm of the cell via evolved receptors in this cells membrane where they are chemically reduced releasing energy pulses at this organs etc operating frequencies and the cell becomes active. Depending on the location and its exposure the cell may react to a mix of chemical (secondary) or primary (direct) energy inputs.

NOTES

1. When the environmental energy input pulses are persistently changed the RM of the relevant gene is damaged and the nucleotide pairs are destroyed with the adjacent nucleotide pairs being stressed beyond their elastic limit (The gene) This stressed section now represents the damaging effect of the particular environmental energy pulse and therefore the environment. The RM nucleotide pairs are replaced and added too, resulting in a reactive ability to switch on a signal to the gene. This change to the gene and signal response ability results in an adapted variation to the output of the genes etc and hence the species (Evolution) as the DNA survives (Ref 1.)

2. When the cell has reached the formative pluripotent stage and prior to the membrane receptors forming it may be differentiated, by intervention, at this stage, with exactly the right potential environmental energy effects, that will allow of the cell being adapted to whatever role is required of it.

3. The process of evolution is one where the energy input from the environment to both parents should be consistent otherwise genetic inconsistencies may occur, possibly creating problems that may effect any characteristic of the offspring.

CONCLUSION

Stem cells do not exist. The effects of the so called Stem cells, contrived to exist, because the process is not understood, is due to normal cells, during the process of division, being inadvertently exposed, in the main part, to the energy effects present in the organ, joint etc of the part being treated. Where the cells are exposed to a faulty environmental energy input in the likes of say a Petri dish then problems can arise.

P.S.

The common sense aspects and logic of this situation then is that the Zygote (germ) cell is a single entity, with no sign of a "Stem cell" but there is evidence of the presence of chemicals, proteins etc in the egg yolk, and these chemicals are energy positive and will release this energy in controlled pulses during their chemical reduction as a result of applied heat energy. On the development of the Embryo cells there is still no sign of the physical presence of a "Stem cell", but they begin developing into the various types of differentiated cells due to the random distribution of these chemicals to individual cells.

Logically, as the various types of cells have been evolved with an ability to carry out survival functions then there is a certain amount of inanity about the belief that a cell has evolved to produce further evolved stem cells that then go on to produce further cells to produce a function and then these cells then produce further stem cells etc and so on.

What Rot!

REF 1

"The Theory of Life and Evolution". Unpublished by John A. LeRoy

EVOLUTIONARY DIVERGENCES as a RESULT of CHANGED ENVRONMENTAL HABITAT EXPOSURE

◇◇

Although not immediately apparent the effects of exposure of members of a species to different habitat environments over a period of generations result in evolutionary divergence to the specimens and if reintroduced on rapid artificial migration they reproduce then the offspring can inherit genetic defects as certain genes have diverged as a result of being exposed to different environmental energy pulse effects leading to their survival in the varying environments and as a result have become no longer fully compatible.

Examples of the divergence of human characteristic where different traits have evolved are Caucasian, Chinese, African, Australian Aboriginals etc. and yet we are of the same "race", becoming different varieties that are not necessarily fully compatible and this fact may be illustrated by reference to other species, for example by the rejection of African elephants of Asian elephants, where incompatibility is sensed. In humans this "rejection" may be considered "racism"

All characteristics of any species, including humans, are subjected to responding as depicted by Mendel's Laws of Inheritance and these characteristics include mental behavioral traits as well as physical attributes. Any characteristic defects observed in a member of a family are not necessarily going to appear in all members as the effects in accordance with the laws applying to Dominant genes and Recessive genes may come into play, however once again in accordance with the laws of Inheritance these flaws may reappear in later generations.

The defects can be so severe that infertility may result plus such defects in humans as diabetes, MS etc. and as long as uncontrolled rapid migration and integration, particularly from the more divergent populations, is occurring the situation will worsen. These problems have been illustrated by a statistics analysis depicting the far above average onset of such problems as Diabetes Type 1, MS etc. applying to integrated Multicultural populations such as the US of A, Australia, New Zealand, and the UK etc.

These defects will range across every characteristic of the species as they are all part of the inheritance and evolution process and this can account for much of the aberrant behaviour becoming apparent in multicultural societies.

All species are subject to this problem including bacteria and when exposed to anti- biotics (The anti-biotics becomes part of their environment when on chemical reduction release aberrant energy pulses) the bacteria evolve rapidly and thus survive.

This situation reveals that "Natural Selection" is not the process by which evolution occurs and highlights the fact that it is non-scientific conjecture. (Nor is "Intelligent Design" or "Creationism"), however "racism" can be the result of rejection, as the divergent members of a species sense incoming energy impulses from the other specimen leading to a reaction of rejection because of what are basically divergent genes present in the rejecting specimens that are not fully compatible resulting in the breakdown of "natural selection" which is in fact about survival of the species, not evolution.

Examples illustrating and highlighting the problems are for example that it has been reported by analysis can locate the source of long settled habitants to within 8km's of the family home.

An example of related problems that researchers have identified and reported in the Journal "Nature" is that with climate change and the polar ice melting, different varieties of bears, seals, whales etc. are being brought together and probably in the absence of compatible mates, interrelated reproduction is taking place. Interspecies sex is reported, however this is disputed as it is intervariety sex as the particular species fore bearers all evolved from a recent common ancestor.

The research concludes that the hybridized offspring run the risk of being infertile with the resulting demise of the species and there is some indication this is occurring with the hybridizing of the human species.

Another example of this is some years ago Indian Scientists, carrying out breeding experiments between Indian and Asian lions were forced to carry out vasectomies on scores of the crossbred offspring to let them die out in a controlled extinction as they were suffering from deformations and diseases

Here we have two sub cultures of the one species with genetic variances due to their evolvement over a long sustained period in different environments. When cross breeding takes place some of their genes have diverged sufficiently for problems to occur due to these genes etc. of the DNA not being compatible. This demonstrates members of a species are best suited to being exposed to a similar habitat environment and as divergence of the environment slowly occurs all involved members should be exposed to this environment change for ultimate results.

Does this problem apply to humans? Of course it does, as an animal we are subject to mutation (evolution). When breeding occurs between individuals, who through our migratory habits, have evolved with a variance of DNA characteristics, then we may breed offspring with dysfunctional problems, including physical, personality, mental and behavioural traits.

The available conclusion to us then is that rapid unrestricted migration of the human subspecies forming multicultural communities and interbreeding has serious scientifically supported implications as far as the health of the expanding population is involved.

Further the possibility exists that some recently arrived individuals and hybridized offspring are at risk of suffering allergies as the sustenance normally available is a conduit for environmental effects. This problem may also extend to affecting individuals when the sustenance is imported.

a team of researchers at the University of Edinburgh that genetic tests have revealed that DNA

An inescapably conclusion arising from this is that the energy radiation being emitted from smart meters and other man made devices such as wind turbines, mobile telephones etc. can inflict serious damage, which although not being immediately identified by apparent physical damage to the person has the potential to cause widespread devastation to our civilization in the coming years. (As occurred to the Army personnel exposed at Maralinga to nuclear radiation some 40 years ago, the effects of which were not apparent at the time).

LIFE, EVOLUTION and ENERGY PULSES

This paper examines the phenomenon of life and evolution and using a scientific approach reaches the following conclusions. These are based on the following:

a. Vibration and resonance is the most common physical phenomenon in the universe.

b. All effects have a cause, there is no such thing as a spontaneous action.

c. All living species have DNA molecules and they are the only reactive entity in any cell. (Viruses are a semi life form).

d. DNA molecules are constituted of nucleotide acid pairs with sugar backings forming a ladder like structure. The nucleotide pairs are subject to resonating on the application of vibrating energy.
   (Attach. 1).

e. Atoms of carbon and Potassium are always present in any cell of any species and a proportion of these elements are radioactive isotopes (Carbon (C14) and Potassium (K40))and they provide a magnetic field in all living species.

f. The presence of an electric current in a species cannot occur without being generated

SOME CONCLUSIONS

1. Life is the damage control electro/chemical/mechanistic reaction of a species DNA as the molecule is subjected to energy pulses emanating from and representing all aspects of its environment.

2. This results in every living species being compatible with its particular environment as the evolved DNA delivers a controlled reaction to the damaging effects of the energy pulses.

3. Life affect is the ongoing initiated reactions of a specimen as the incoming environmental energy pulses gain controlled access to the physically (mechanistically installed by vibration) recorded effects of previous similar exposures of the DNA (The genes). There is an electro/chemical/mechanistic process by which these records are handed down from generation to generation and this is the inheritance phenomenon.

4. In the more complex species capable of controlled motion a mental capacity has evolved utilizing gene responses to the complex RM (Regulatory Memory) of Glia cells to access recorded information (Recognizable physical distortions of the relevant evolved nucleotide pairs of the DNA cognizant with the frequency of the incoming energy pulses from the various senses) to initiate a physical reaction. This capability (Mental and logic characteristic) was initially established enabling a physical control capacity for sustenance gathering and therefore resulting in survival.

5. Species are evolved as they are exposed to the developing environment and hence their DNA to the relevant energy pulses, causing a survival reaction by the cell. They are adapted to the changing environment when a persistent exposure to changed environmental energy of a limited nature occurs with a resultant gene (DNA) adaptation (Evolution) and the more complex the environment affecting the species the more complex the species.

6. The radiation (Energy pulses) emitted by manmade sources, such as smart meters and mobile telephones (Wind turbines also emit energy pulses we are not evolved to withstand) result in exposures that are significantly different to the normal environmental exposure experienced and the change is so rapid it causes irretrievable damage to the DNA when the specimen

is persistently exposed. This effect can cause illnesses such as cancer over time and has the potential over an extended time frame to eventually devastate exposed populations.

7. Understanding the life/evolution process leads to ability to establish the causes of genetic problems such as cancer, diabetes, multiple sclerosis, mental problems such as dementia etc. and autism, in fact an understanding of the basis of the life process can lead to the resolution of the causes of all genetic problems.

Life/Evolution and Energy Pulses. (Both Environmental and Non environmental)

In the beginning an arid, hot, barren world came into being. As eons of time passed and cooling occurred various elements, including Carbon (C) were being dispensed.

These elements were all chemically active particularly the Carbon that due to its nuclear arrangement its atom had an affinity to form numbers of chemical molecules incorporating the same numbers of atoms of various elements bonded together in different arrangements (Isomer molecules). These may have different energy levels bonding the atoms of the molecule together.

Eventually with the cooling of the planet, water molecules in liquid form became available as did molecules of sugar (chemically formed, naturally) along with other elements including Phosphorous (P) Nitrogen (N) Hydrogen (H) Oxygen (O) and Carbon (C). Also in this swampy background were various readily available elements including Potassium (K) incorporating radioactive isotopes (K40) that along with radioactive carbon (C14) formed a magnetic field. (Every living cell has Potassium incorporated into it or a radioactive alternative).

Over time complex RNA like molecules formed with molecules of nucleotide acids A, C, G, and T bonded into a backing of a carbon based molecule (sugar) with its affinity to incorporate other elements.

These molecules of acids with their sugar backing became further locked together with the extremities of the sugar bonding to Phosphorous atoms. The process developed with the molecules of the nucleotide acids having chemically reactive free protrusions that have an affinity for bonding together via hydrogen atoms in the order of acids A to T and G to C with the potential to form nucleotide pairs.

Eventually the statistical possibility occurred and short lengths of RNA like molecules matched up in order with the appropriate pairings and the molecule DNA appeared. With the sugar extrusions bonded in a continuous length and the nucleotide acid protrusions cross bonded the complex molecule resembled a ladder like structure.

The format of this molecule was such the nucleotide pairs initially vibrated when exposed to the photons constituting the environmental heat energy pulses (Attach. 1). (The phenomenon of Vibration and resonance is the most common in the universe). (All effects of any environment on the DNA are by way of energy pulses delivered by either primary or secondary means and they cause a reaction from it). (There is always environmental energy present at temperatures above -273°C (0°K)

The vibration caused a reaction from the molecule when destructive damage to the nucleotide pairs not in resonance with the frequency of the energy pulse resulted in the chemical breakdown of the out of tune pairs, forming an enzyme that assisted in a hit and miss chemical response producing additional and replacement pairs that eventually developed into a DNA molecule with a stable harmonic arrangement reactive to the vibrating energy pulses.

As these nucleotide pairs resonated in reaction to the energy pulse they cut through the magnetic field, provided mostly by the C14 and K40 radioactive atoms, generating an electrical current in them with an induced surrounding force field that rose and fell in reaction to the initiating energy pulse that was representative of the environment. This energy field catalyzed a chemical reaction that consisted of a hit and miss production of amino acid molecules held together by peptide bonds (Protein) which was directly in contact with the DNA and intervening between it and the energy pulse (Wrapped, scab like). This process, building the protein (Chromatin, Epigenetic effect), eventually stabilized when the incoming energy pulse was muffled to such a degree by the adjusted amino acid molecules arrangement that the energy pulse effecting the DNA molecule was sufficient only to maintain the effectiveness of the protein as it provided protection for the DNA from the damaging destructive effects of the relevant energy pulses.(DNA is the life giving entity whilst the Proteins etc. are the building blocks of life).

During this process excess protein formed around the molecule and protein forming a membrane creating a self-contained environment (Cell forerunner, an Autotroph, this cell reacted directly in response to environmental energy, reducing, as part of its process, carbon dioxide and producing sugar as it released Oxygen. There is a cause and an effect, no cell is spontaneous in function) (The DNA is the only reactive molecule in a cell, and is present in every living cell and is therefore the life entity).

The makeup of the protein molecule was governed by a hit and miss replacement process of the nucleotide pairs until the resultant amino acid molecule disposition resulted in a muffling effect to the particular frequency (Attach. 2), reducing the impact of the energy pulse on the DNA nucleotide pairs to a stable level resulting in extended survival of the DNA molecule until due to persistent exposure to the reduced effect of this pulsing energy the nucleotide pair suffered material fatigue and ruptured at the hydrogen joints. With the chemical characteristic being present to catalyze chemical replacement of the missing DNA half it resulted in the two divided halves of the molecule forming two DNA molecules (cell propagation)

Eventually a slight but persistent change occurred to the DNA's environment to which it was exposed and hence a slightly changed energy pulse of a frequency to that being encountered directly penetrated the protein gaining access to the DNA (Attach. 5.). Damaging several relevant nucleotide pairs to the point of destruction these were chemically replaced and added to in a hit and miss sequence until they were in a harmonically reactive arrangement to this additional frequency whereupon the damage and repair procedure ceased. These nucleotide pairs were now reactive to this particular frequency and resonated relatively to the strength of the incoming changed energy pulse as it made its way through the muffling protein. (These reactive nucleotide pairs are known as the Regulatory Memory (RM) and have been identified by Biologists as "switching" the genes on).

(Genes were/are formed as a result of the damaging effect of this changed energy pulse on the relevant nucleotide pairs adjacent to those destructively damaged that were replaced, and added to. As they were strained beyond their elastic limit but not destroyed they (the Gene) now represented the effect of this potentially damaging aspect of the original environmental energy pulse and were a permanent record (memory) passed on by the propagation and eventually inheritance process).

The initial process of the propagation of the DNA is electro chemical and the installation of the genes that indirectly react to physical inputs (Energy pulses) is a physical process and there is an interval between these two events occurring. This is now known as the pluripotent stage of the cell. Later in the history of evolution of life with the event of the evolution of differentiated cells, it resulted in the complex species survival and this was achieved by the specific genes of the cells reacting to specific energy inputs. The application of these controlled relevant energy inputs occurs at the pluripotent stage, establishing a specific output from the common DNA molecule and established differentiated cells with outer membranes with receptors that accept the specific chemicals which are conduits for environmental energy pulses and therefore the environment that on controlled reduction within the cell release specific energy pulses that cause a reaction from the DNA relevant to the survival of the species.

These reactive nucleotide pairs (RM) generate an electric current (contained within the DNA length inclusive of these nucleotide pairs by a property known as "Quantum tunneling" (Attach. 3) and therefore restricting the switching effect to the relevant activated gene) as they were resonating in response to this energy pulse, causing a force field to be initiated that reflected this aspect of the environment and as it rose and fell it "switched" the sensitized gene nucleotides to catalyze a chemical molecule reaction (RNA) that initiated the reinforcement of the protein leading to the protection of the DNA from the damaging effects of the energy pulse and hence the single cell "Prokaryote" species was compatible with its environment.

As the environment underwent further changes the operation was repeated with extension of the DNA molecule as further genes and RM's were formed with multiple specimen variations occurring as they were subjected to changed environmental exposure effects. As gene change occurred the next variation, of necessity, supported the previous causing the DNA and its evolving specimen and hence multiple specimens (the species) to survive (live) in a particular environment. (The established genes were passed on from one generation to the next as the nucleotide of the existing half of the gene was strained beyond its elastic limit and on the other half being replaced when the vibrating energy pulse was applied the inherent weakness of the original nucleotide led to the rapid weakening of the new half, reinstating the gene. Once installed in a species DNA a gene is permanent, whether functional or not).

The RM and gene reacts to an energy pulse over and above that of the normal incoming heat energy and where a gene is activated in a cell/differentiated cell the invigorated reaction results in the formation of a clump of protein around which the relevant gene etc. is entwined (Histone) plus overflow into the chromatin. This clump of protein thrusts the gene clear of the chromatin protein making it available for RNA replication (Attach. 4).

The result is the observed differences in the chromatin is erroneously thought to code for characteristics as does the DNA whereas it is actually muffling the incoming environmental energy pulses with frequencies relevant to the differentiated cell, protecting the DNA from extensive damage and therefore leading to survival of the species.

Due to the writhing motion caused by the energy pulses the extended DNA molecule ends eventually joined forming the DNA circular format of most of the prokaryote species now in existence.

This is the basic process of all formats of life, obscured in certain species as they have evolved into more complex forms of the various Eukaryotic species as their environments have developed. The basic principle is in all cases the DNA reacts to energy pulses that are directly attributable to its environment and it has been evolved to resist the damaging effects of these energy pulses and as persistent slight environment change occurs the DNA reacts and is evolved making it and the specimen compatible with the environment and thus survival of the DNA. (As environment effects are delivered chaotically and the DNA are extremely sensitive to change the conscious life characteristic is a reaction to chaotically varying changes and is present in varying degrees depending on the evolutionary development of a species as its DNA reacts to its environment).

Some two billion years ago a prolific single cell species had evolved that had the reaction of producing sugar via the sun excitation/ chlorophyll process e.g. lichen etc. (The sugar was/is a conduit for environmental heat energy). In the presence of this process a cell evolved with the capacity to produce energy by chemically reducing sugar. During this period other single cell bacteria like species evolved that directly consumed sugar from the single cell lichen species and indirectly released sugar molecules. With the input of energy via the sugar there was an output from these cells in the form of propionic acid and this acid was an indication of the presence of and an inciter for the evolving species to absorb sugar when in the presence of this output.

This is the Mitochondria cell species and eventually it invaded a host prokaryotic cell (symbiotic relationship) with the energy pulses it released activating the DNA of the host cell leading to a more vibrant and active reaction from the cell over the full 24 hr. period. Eventually a further similar single cell invaded this infected cell and with its increased energy utilization the invaded cell was capable of supporting the DNA of the secondary cell as well as its own. This was the process involved in the establishment of the combination "Eukaryotic" cell and the activating process for this complex cell was the forerunner of the sexual act.

During this period of rapid change the eukaryotic cell evolved sophisticated mechanisms including the protein processing ribosomes, organelles etc., with the nucleus and the cytoplasm being effectively reliant on control from the DNA as it reacts to the environment influence. Additionally the circular DNA molecule underwent strain to such an extent enclosed in the initial cell casing (the nucleus) it ruptured into various lengths forming chromosomes encased in the protective chromatin protein.

If a mutative effect occurs caused by radiation, virus infection, asbestos etc. to the DNA nucleotide pairs that is beyond its renewal capacity the protective protein output formulae is disrupted (Chromatin, as the DNA alphabet or code for the production of protein is changed) and incoming environmental energy pulses can then penetrate to the damaged DNA resulting in further rupture of the nucleotide pairs and rapid fire propagation of the cells with damaged DNA and dysfunctional performance in duplication of cells as they attempted to build parasitical organs (Cancer).

(From point A, the beginning of life phenomenon to point B, the present, it seems some forms of cells and their outputs are extremely complicated and unexplainable, however with an understanding of the basic process and an appreciation of incremental changes over the four billion years time span the mystery unfolds)).

Gradual increasing exposure to different conditions (environments) various cells underwent processes of adaptation (evolution) as the relevant energy pulses drew reactions and development of genes from the two sets of DNA resulting in their survival with individual output s controlled by genes being compatible. A major requirement for survival was the exposure to sustenance (sugar) to support its energy requirements and various methods dependent on circumstances developed e.g. plants extract requirements from the soil and air whilst being energized by the sun excitation/chlorophyll process and forerunners of fish species drifted to their sustenance obviating to a large degree the need for movement control, and direction to for acquisition of sustenance therefore evolution of extensive logic and brain power.

Other Eukaryotic cells were evolving along different processes depending on environmental exposure and this included a species that eventually evolved into the mammals.

For this developing species exposed to damage from the increasing environmental effects the outputs from the cells of necessity meant an increase in energy necessary for survival. Exposed to an environment where it was subjected to the propionic acid isomer molecules from the bacteria that incited the recognition of the presence of and absorption of the sugar the increased energy availability resulted in the multi cell species surviving. As the species environment further developed it gradually evolved, increasing the numbers of multiple cells reacting to increasingly different energy pulses delivered from primary and developing internal chemical means (secondary) in support of the preceding cells (differentiation) and becoming more complex and during this developmental stage the supporting bacteria was incorporated into its makeup resulting in mammal species dependent on a gut flora symbiotic relationship.

This symbiotic relationship persisted and does so today aiding in the recognition and the mental outputs to acquire sustenance with its propionic acid molecules output. Without this output, during the conception stage to the embryonic stage when the cells acquire appropriate receptors there would be no access to the cells and therefore to sustenance. When the umbilical cord attaches to the fetus the mother supplies the variety of chemicals required for the developing differentiated cells until the time of birth, where outside environmental influences take over its development until the child reaches maturity. With this system the mother's health and practices can influences the child's development over its lifetime (e.g. if mother is placed on a diet in first semester or suffers from lack of sustenance her chemical reaction, passed on by chemical means can result in child developing cells that instigate behavior resulting in compulsory eating and therefore obesity. Without this symbiotic relationship mammals would not exist, however the advantages outweighed the negatives as the process resulted in an efficient access to sugar that supported their development.

As the evolving species requirements to access survival sustenance increased the senses along with limbs were evolved in support of mobility with cells being established with the capacity to build flesh, bones etc. allowing access to sugar and hence survival. Further it necessitated the characteristic of logic and reasoning to acquire this sustenance and cells evolved in support of an existing communication system between the cells (Eventually lead neuronal cells of the brain). (The DNA of each and every cell and therefore the cell reacts selectively with an output due to external energy pulses (the input) in a mechanistic manner as physically induced reactive distortions of the DNA nucleotide pairs of various descriptions are installed (The RM and genes) that are recognized and interpreted leading to reactivation of the cells by ongoing inputs and therefore outputs resulting in survival of the species. As the environmental input changes the output capacity requirement results in DNA molecules dedicated to a task and additional cells in support of the previous cell when an increasingly complex physical output for survival is required.

When a mental (conscious and unconscious) response is required capacity to store information in an accessible mechanistic format (memory) enabling a logical (accessing and sorting) reaction is required to enable survival the capacity is provided by extension of the memory (RM) portion of the DNA molecule (Junk) and an increase in number of supporting "Glia" cells. Some of these evolved in support of and dedicated to various tasks such as the senses (consciousness) whereas others evolved dedicated to subconscious mental outputs such as balance control in walking. As the environment affecting each sense increased becoming more complex additional supporting cells were evolved that had the capacity to react to the increasing information and relay it selectively in funnel like fashion to the relevant neuronal cell containing the relevant memory, if any. Inputs from the various senses were coordinated as they evolved, leading back sequentially to the cell with the initial output source, capable of causing a physical reaction.

Repeated persistent exposure to a subject caused mechanistic distortions of the DNA nucleotide pairs representing a recognizable input and depending on the degree of exposure and distortion these strained nucleotide pairs recover at a variable rate. With restricted exposure they eventually recover completely, but with repeated persistent exposure they become virtually permanently distorted i.e. long term memory and if the total inputs are between these limits graduated memory terms occur. If the incoming information alerts a memory and it requires a response an action output is initiated. If an action is seen as spontaneous such as suckling it is due to the specimen's ancestors having been subjected to the associated stress repeatedly and the inherited junk nucleotide pairs permanently distorted. Additionally to convey these inputs to the memory centres, synaptic connections between the glial cells are injected with a protein capable of conducting the appropriate signals but stifling others (Attach. 6). If this protein is not reinforced during further controlled inputs allowing quick reflex actions, it is destroyed and the components recirculated by maintenance cells. With the onset of ageing these maintenance cells may deteriorate and there is an uncontrolled buildup of the protein chocking and killing of cells resulting in Dementia i.e. long term memories still available to a degree but short term badly affected.

This line of species in time diverged as different groups were exposed to slight persistent environment changes trending to evolve them into numerous species as they that became divergent. Due to the necessity to support this change it lead to the adaptation of further characteristics, resulting in different sustenance acquisition demands and consequently mental control and physical abilities of various complexities. Mankind owes its existence and mental capacity to the complex environments over the course of history it has been exposed to.

This accounts for many of the earlier evolved genes having similar outputs but genes with variations as the species diverged. It also leads to the inescapable conclusion that all characteristics including mental ability and capacity are installed by a mechanistic physical process as a result of exposure to the ongoing nurturing environment of a species and nature is a result of the selectively accessed recorded effects of the reactions by the DNA to the specimen's nurturing over generations. (Mendel's Laws of Inheritance). Therefore mental reactions cannot be entirely controlled by an individual.

Further conclusions

1. Cells do not and cannot spontaneously propagate. (Law of Physics, Cause and effect)
2. Specimens with parents from divergent environments may inherit incompatible genes resulting in serious genetic problems e.g. Diabetes type 1. MS. (Attach. 7).
3. Exposure to persistent energy pulses outside of those normally encountered in a species environment leads to extensive genetic damage with resultant problems such as cancer, because Evolution has not evolved protein with arrangement of amino acids molecules to protect against the man made energy pulses.
4. Species do not adapt to their environment, their DNA molecules are activated and react to the environmental energy effects, otherwise they would not be in existence.
5. Darwin's Theory of Evolution that evolution occurs by natural selection does not hold up in that it means, it is a spontaneous process and by the laws of physics every effect must have a cause.
6. Stem cells do not exist. The effect is achieved when cells with common DNA molecules are activated, during their pluripotent stage as they propagate, by the correct application of the applicable energy pulse required to activate the pertinent effects.
7. Clostridia bacteria, part of the gut flora emits Propionic acid molecules as its output that ingress brain cells in their role of assisting in identifying sustenance and aiding in its acquisition. This bacteria is physically passed on from mother to child and takes three years in the child to become stabilized. The acid molecules. Carbon based, are isometric i.e. may have different energy content depending on their structure as a result of the state of the evolution of the bacteria. Bacteria as evidenced by recent developments are prone to rapid evolution and antibiotics when administered during their stabilization period is part of their environment and this results in the output production of a changed evolved isometric molecule. When chemically reduced in the brain cells they release energy signals that are different and confusing resulting in dysfunctional physical and mental

reactions (Autism). Drastic change of sustenance (diet) to that normally experienced by a migrant mother can also induce the problem in a young child as the bacteria has evolved to deal with the mothers initial diet.

Attachments

1. "Introduction Wave Genetics, Genes, DNA" etc. Extract from http://www.soaceandmotion.com/evolution-biology-wave-genetics.htm
2. Extract from "science of Zoology" Mc Graw-Hill Book Co. Penetration of Protein molecule by energy pulses.
3. "Quantum Tunneling" Extract from http://www.newscientist.com/article/dn-electrifying-claims-for-dna-are-dashed.html
4. "Histones" Scientific American Dec 2003.
5. a. Low intensity microwave radiation as modulator etc. by Vojisavljevic V. Pirogova E. Cosic I.
   b. Demodulation in tissue, the relevant parameters and the implications for limiting exposure. By J. Silny.
   c. DNA fragmentation in human fibroblasts under extremely low frequency electromagnetic field exposure by Focke F. Scherman D. Kuster N. Schar P.
   (Papers demonstrating the phenomenon of non environmental energy pulses bypassing protein and damaging DNA).

6. Extract from "Biology Today" Random house publishing Extract is "The Molecules of Life" (Protein).
7. Evolutionary divergences resulting from Environmental Exposure.

Purpose and substance of each enclosure

1. As a result of experimentation demonstrates the DNA is reactive to environmental energy impulses

2. Demonstrates the layout of the amino acid molecules of the complex protein molecule and their symmetry with format of pulsating energy and what it means in respect of cancer.

3. Establishes how the generation of electricity is confined to a restricted length of the DNA resulting in a controlled signaling or switching of the gene

4. Clearing of the gene from the clutter allows efficient production of the RNA.

5. Experimental results establishing that contrary to authoritative views that LF radiations such as emitted from smart meters do not represent a danger to human populations, they do.

6. Indicates that Protein in any of its many forms is utilized by the cells to undertake functions leading to its and hence its species survival.

7. Demonstrates just how sensitive DNA is to exposure to environmental habitats.

FUNCTION OF THE AUTO-IMMUNE SYSTEM

◇◇

Many of the findings and conclusions, as seen through the eyes of a professional engineer, are outside of the concepts that biologists and geneticists have as they are based on engineering principles, physical phenomena and recognized biological experimental results.

The following is based on the understanding of the life process put together over many years and it does not contain all of the relevant details necessary to completely gather a full working knowledge of such items as for instance, what is the basic system that drives the functioning of a cell. (They are not and cannot be self-propagating, according to the basic laws of physics of Cause and Effect). Many more findings are contained in the attached papers.

Having read an article re work concerning the Auto Immune system in the Sunday Herald Sun and looked up the website, I find myself puzzled as to how some obvious facts appear to play no part in the research of many and I would be interested as to how the conclusions were arrived at.

First it is a fact that the only common active entity of all species is DNA of similar molecular formats. These molecules consist of chemicals made up of the elements C, H, N, O and P. that are in themselves inactive.

Secondly when a species and hence its DNA molecules are gradually, over extended periods of time, exposed to different environments, variations to the genes of the DNA molecule occur and eventually reactions from differentiated cells result and the adaptation of the species characteristics to this environment takes place. e.g. Caucasians, Africans, Chinese, Australian Aboriginals etc. i.e. varieties, not different race species.

As such when these elements are formed into DNA molecules they are chemical molecules that would be passive, unless activated by an outside influence. This of course also entails the installation of the characteristics such as genes etc and this cannot happen without an outside influence being applied. (There is no mystic influence, nor can there be).

The fact then is that the only form any outside influence can take is that of energy pulses and the molecules of DNA have been demonstrated to be susceptible to the energy pulses of the environment that cause the nucleotide pairs to vibrate (activated) (Prof Pjotr Garjajev and Russian research team). Once the variations of environmental energy inputs occur, supporting cells (differentiated) are developed with output reactions resulting from specific genes etc leading to survival of the DNA (and species) in this environment and from passive materials "life" eventuates as chaotic application of reaction initiating energy pulses occur.

Note

Species, based on the only common reactive entity that is DNA, a chemical molecule incorporating passive elements, do not and cannot respond to the environment, the effects of which are always passed on to them by energy pulses, but they react to it. When, as conceived by biologists, if responding to its environment, life outputs from the DNA could not continue and would not commence i.e. what would be responding?

It has also been factually established that variations to the DNA molecules i.e. genes etc occur depending on the environment the specimens are exposed to and this includes specimens of the same species i.e. they have as yet not been exposed to sufficient variation of environmental energy impulses to evolve them into different species.

Each and every species, based on its DNA is compatible with the influences and effects of the environment which are conveyed to it by the various energy pulses emanating from it and the DNA molecule then reacts resulting in the chemical production of proteins hormones etc the build up of which protects it from further damage against these specific energy pulses and further provides for the species survival. As permanent changes of the environment occur reactions to the variation of energy inputs occur that cause the adaptation of the DNA molecule with a modified output that results in changes to the species and its survival (Evolution). This entails the development of further differentiated cells that are integrated with and support the previously evolved cells in the survival of the species and hence the DNA.

From the above it can be deduced that information (The results of Reactions) leading to the survival of the species is installed in the DNA due to its exposure to the environmental energy and as a result the DNA molecules reacts to any continuing exposure to these particular energy pulses.

The process of the DNA reacting then induces a chemical process, protein's, hormones etc. that can be continuing and repetitive and this is seen as a cell having the property of memory, which it is not, it is the DNA of the cell having the permanently induced potential of reacting to particular activating energy inputs.

With bacteria and virus cells with the capacity to invade and feed off normal cells for reproduction purposes thereby taking advantage of the cells energy producing characteristics the infected cells DNA molecules may be damaged and consequently have an output of protein etc., the antigen, not compatible with its evolved purpose.

The auto immunology process evolved ensuring survival of the DNA. Given some cells are initially infected a process evolved was differentiated cells with an output capable of containing the damage, (T and B cells) with the potential to produce a protein (The Antibody) that has a design suitable to assimilating or destroying the infected/ damaged cell producing the antigen molecules.

Initially this antigen molecule(s) with the onset of the disease accesses the antibody-producing cell which then becomes more specifically differentiated to produce a particular antibody protein and on the chemical reduction of the antigen within the cell, energy pulses are released that activate the "Regulatory Memory" of the gene(s) that are responsible for activating the genetic system of producing the antibody protein. This is in accordance with the process of life and evolution, where each type of cell has been evolved to complement the previously evolved cells when a variation to the energy input occurs resulting in the eventual survival of the DNA molecules. The T and B cells in this process are then available to intercept the defective protein molecules (Antigen) produced by ailing cells and as a result produce antibody protein molecules that integrate with antigen molecules that are continually being produced by the infected cells and by one of several means such as muffling, the (these) ailing cell(s) is (are) then negated.

Note

While the DNA molecule is the reacting entity that identifies and defines the process to be undertaken the protein, hormones etc are produced and utilized to facilitate the various functions required to achieve survival of the DNA molecules in the damaging effects of the energy pulses associated with its chaotically varying environment, where the DNA molecule is rapidly and continually selectively reacting. The process is such that the course of evolution has ensured the protein output etc exactly matches and facilitates the requirement of the specific DNA molecule to survive in the effects of the environmental energy inputs. (This is the reason transplants are normally rejected) The species can therefore be seen as virtually a scab effect protecting the DNA molecule. Protein in itself cannot be seen to initiate responses as some times depicted by biologist and this includes the epigenetic effect.

For Eukaryotic species specific energy inputs in support of the direct environmental input are derived from a secondary means (Food sustenance) by the importing of relevant environmental energy pulses (e.g. a sugar molecule in the controlled chemical reduction process in a cell releases approx. 130 pulses) and this "means" is by molecules of protein, hormones etc, produced in a cascading regime initiated by the environment, accessing differentiated cells in the order of their evolution. In the case of the differentiated T cell it has evolved to produce an antibody initiated by the ingress of the initial specific antigen being produced by the infected cell with the protein antibodies produced having the structure to counter further antigen molecules leading to destruction of the infected cells and overall survival.

Note

The infecting agent is interpreted as an environmental effect, (Like all species it is a being of the environment) with the release of energy pulses within the cell and thus initiating a reaction from the DNA molecule resulting in the production of the antigen protein by the cell.

This then demonstrates that the concept of "killer" and "memory" cells is questionable as the whole process including the auto immune system is one of a very complex mechanistic, electro chemical process and is purely dependent on the reaction of the evolved cells DNA molecules to environmental effects when eventually exposed to the antibody initiating energy pulses from the antigen.

Terms such as "immunes system's ability to identify, remember and kill invading cancer cells" and various other statements implying some sort of thinking and self initiating response of the cells is an ability that does not exist and leads to an extremely misleading state of confusion among researchers. These reactions, as previously stated, are purely electro chemical processes and occur mechanistically.

Cancer cells are inherently more difficult for the auto immune system to cope with as with the chromosomes being chaotically ruptured as the cancer cells develop, the immunity generating cells cannot cope with the rapid variation of the protein being produced. A more likely treatment is doses of olive oil, which being rich in monounsaturated fat molecules results in the molecules being converted to mega molecules of chylomicron in the lower intestine which are transported by the blood, and as most cancer cells are reacting at fifteen to twenty times the rate of a normal cell the energy carrying blood is drawn to them and the chylomicron molecule with it smothering capacity prevents the energy process becoming available to the cancer cell resulting in hopefully their demise. This process has the potential to be successful with all of the affected cells including secondary cancers as the blood with the energy providing sugar is drawn to each of them and is possibly the reason that statistically, Mediterranean people suffer fewer cancers than other populations.

The remaining papers fill in the details of the complete process. The papers as previously stated are based on recognized physical phenomena, biological experimental findings, electrical and materials engineering processes. The findings integrate together to such an extent it is virtually impossible to logically fault the conclusions, however they do highlight the belief in dogma that biologist accept unquestioningly are holding up the work of establishing the causes of many of the genetic problems facing the world today.

LIFE, EVOLUTION and ENERGY PULSES

◇◇◇

This paper examines the phenomenon of life and evolution and using a scientific approach reaches the following conclusions. These are based on the following:

a. Vibration and resonance is the most common physical phenomenon in the universe.
b. All effects have a cause, there is no such thing as a spontaneous action.
c. All living species have DNA molecules and they are the only reactive entity in any cell. (Viruses are a semi life form).
d. DNA molecules are constituted of nucleotide acid pairs with sugar backings forming a ladder like structure. The nucleotide pairs are subject to resonating on the application of vibrating energy.
(Attach. 1).
e. Atoms of carbon and Potassium are always present in any cell of any species and a proportion of these elements are radioactive isotopes (Carbon (C14) and Potassium (K40)) and they provide a magnetic field in all living species.
f. The presence of an electric current in a species cannot occur without being generated

SOME CONCLUSIONS

1. Life is the damage control electro/chemical/mechanistic reaction of a species DNA as the molecule is subjected to energy pulses emanating from and representing all aspects of its environment.

2. This results in every living species being compatible with its particular environment as the evolved DNA delivers a controlled reaction to the damaging effects of the energy pulses.

3. Life affect is the ongoing initiated reactions of a specimen as the incoming environmental energy pulses gain controlled access to the physically (mechanistically installed by vibration) recorded effects of previous similar exposures of the DNA (The genes). There is an electro/chemical/mechanistic process by which these records are handed down from generation to generation and this is the inheritance phenomenon.

4. In the more complex species capable of controlled motion a mental capacity has evolved utilizing gene responses to the complex RM (Regulatory Memory) of Glia cells to access recorded information (Recognizable physical distortions of the relevant evolved nucleotide pairs of the DNA cognizant with the frequency of the incoming energy pulses from the various senses) to initiate a physical reaction. This capability (Mental and logic characteristic) was initially established enabling a physical control capacity for sustenance gathering and therefore resulting in survival.

5. Species are evolved as they are exposed to the developing environment and hence their DNA to the relevant energy pulses, causing a survival reaction by the cell. They are adapted to the changing environment when a persistent exposure to changed environmental energy of a limited nature occurs with a resultant gene (DNA) adaptation (Evolution) and the more complex the environment affecting the species the more complex the species.

6. The radiation (Energy pulses) emitted by manmade sources, such as smart meters and mobile telephones (Wind turbines also emit energy pulses we are not evolved to withstand) result in exposures that are significantly different to the normal environmental exposure experienced and the change is so rapid it causes irretrievable damage to the DNA when the specimen

is persistently exposed. This effect can cause illnesses such as cancer over time and has the potential over an extended time frame to eventually devastate exposed populations.

7. Understanding the life/evolution process leads to ability to establish the causes of genetic problems such as cancer, diabetes, multiple sclerosis, mental problems such as dementia etc. and autism, in fact an understanding of the basis of the life process can lead to the resolution of the causes of all genetic problems.

Life/Evolution and Energy Pulses. (Both Environmental and Non environmental)

In the beginning an arid, hot, barren world came into being. As eons of time passed and cooling occurred various elements, including Carbon (C) were being dispensed.

These elements were all chemically active particularly the Carbon that due to its nuclear arrangement its atom had an affinity to form numbers of chemical molecules incorporating the same numbers of atoms of various elements bonded together in different arrangements (Isomer molecules). These may have different energy levels bonding the atoms of the molecule together.

Eventually with the cooling of the planet, water molecules in liquid form became available as did molecules of sugar (chemically formed, naturally) along with other elements including Phosphorous (P) Nitrogen (N) Hydrogen (H) Oxygen (O) and Carbon (C). Also in this swampy background were various readily available elements including Potassium (K) incorporating radioactive isotopes (K40) that along with radioactive carbon (C14) formed a magnetic field. (Every living cell has Potassium incorporated into it or a radioactive alternative).

Over time complex RNA like molecules formed with molecules of nucleotide acids A, C, G, and T bonded into a backing of a carbon based molecule (sugar) with its affinity to incorporate other elements.

These molecules of acids with their sugar backing became further locked together with the extremities of the sugar bonding to Phosphorous atoms. The process developed with the molecules of the nucleotide acids having chemically reactive free protrusions that have an affinity for bonding together via hydrogen atoms in the order of acids A to T and G to C with the potential to form nucleotide pairs.

Eventually the statistical possibility occurred and short lengths of RNA like molecules matched up in order with the appropriate pairings and the molecule DNA appeared. With the sugar extrusions bonded in a continuous length and the nucleotide acid protrusions cross bonded the complex molecule resembled a ladder like structure.

The format of this molecule was such the nucleotide pairs initially vibrated when exposed to the photons constituting the environmental heat energy pulses (Attach. 1). (The phenomenon of Vibration and resonance is the most common in the universe). (All effects of any environment on the DNA are by way of energy pulses delivered by either primary or secondary means and they cause a reaction from it). (There is always environmental energy present at temperatures above -273°C (0°K)

The vibration caused a reaction from the molecule when destructive damage to the nucleotide pairs not in resonance with the frequency of the energy pulse resulted in the chemical breakdown of the out of tune pairs, forming an enzyme that assisted in a hit and miss chemical response producing additional and replacement pairs that eventually developed into a DNA molecule with a stable harmonic arrangement reactive to the vibrating energy pulses.

As these nucleotide pairs resonated in reaction to the energy pulse they cut through the magnetic field, provided mostly by the C14 and K40 radioactive atoms, generating an electrical current in them with an induced surrounding force field that rose and fell in reaction to the initiating energy pulse that was representative of the environment. This energy field catalyzed a chemical reaction that consisted of a hit and miss production of amino acid molecules held together by peptide bonds (Protein) which was directly in contact with the DNA and intervening between it and the energy pulse (Wrapped, scab like). This process, building the protein (Chromatin, Epigenetic effect), eventually stabilized when the incoming energy pulse was muffled to such a degree by the adjusted amino acid molecules arrangement that the energy pulse effecting the DNA molecule was sufficient only to maintain the effectiveness of the protein as it provided protection for the DNA from the damaging destructive effects of the relevant energy pulses.(DNA is the life giving entity whilst the Proteins etc. are the building blocks of life).

During this process excess protein formed around the molecule and protein forming a membrane creating a self-contained environment (Cell forerunner, an Autotroph, this cell reacted directly in response to environmental energy, reducing, as part of its process, carbon dioxide and producing sugar as it released Oxygen. There is a cause and an effect, no cell is spontaneous in function) (The DNA is the only reactive molecule in a cell, and is present in every living cell and is therefore the life entity).

The makeup of the protein molecule was governed by a hit and miss replacement process of the nucleotide pairs until the resultant amino acid molecule disposition resulted in a muffling effect to the particular frequency (Attach. 2), reducing the impact of the energy pulse on the DNA nucleotide pairs to a stable level resulting in extended survival of the DNA molecule until due to persistent exposure to the reduced effect of this pulsing energy the nucleotide pair suffered material fatigue and ruptured at the hydrogen joints. With the chemical characteristic being present to catalyze chemical replacement of the missing DNA half it resulted in the two divided halves of the molecule forming two DNA molecules (cell propagation)

Eventually a slight but persistent change occurred to the DNA's environment to which it was exposed and hence a slightly changed energy pulse of a frequency to that being encountered directly penetrated the protein gaining access to the DNA (Attach. 5.). Damaging several relevant nucleotide pairs to the point of destruction these were chemically replaced and added to in a hit and miss sequence until they were in a harmonically reactive arrangement to this additional frequency whereupon the damage and repair procedure ceased. These nucleotide pairs were now reactive to this particular frequency and resonated relatively to the strength of the incoming changed energy pulse as it made its way through the muffling protein. (These reactive nucleotide pairs are known as the Regulatory Memory (RM) and have been identified by Biologists as "switching" the genes on).

(Genes were/are formed as a result of the damaging effect of this changed energy pulse on the relevant nucleotide pairs adjacent to those destructively damaged that were replaced, and added to. As they were strained beyond their elastic limit but not destroyed they (the Gene) now represented the effect of this potentially damaging aspect of the original environmental energy pulse and were a permanent record (memory) passed on by the propagation and eventually inheritance process).

The initial process of the propagation of the DNA is electro chemical and the installation of the genes that indirectly react to physical inputs (Energy pulses) is a physical process and there is an interval between these two events occurring. This is now known as the pluripotent stage of the cell. Later in the history of evolution of life with the event of the evolution of differentiated cells, it resulted in the complex species survival and this was achieved by the specific genes of the cells reacting to specific energy inputs. The application of these controlled relevant energy inputs occurs at the pluripotent stage, establishing a specific output from the common DNA molecule and established differentiated cells with outer membranes with receptors that accept the specific chemicals which are conduits for environmental energy pulses and therefore the environment that on controlled reduction within the cell release specific energy pulses that cause a reaction from the DNA relevant to the survival of the species.

These reactive nucleotide pairs (RM) generate an electric current (contained within the DNA length inclusive of these nucleotide pairs by a property known as "Quantum tunneling" (Attach. 3) and therefore restricting the switching effect to the relevant activated gene) as they were resonating in response to this energy pulse, causing a force field to be initiated that reflected this aspect of the environment and as it rose and fell it "switched" the sensitized gene nucleotides to catalyze a chemical molecule reaction (RNA) that initiated the reinforcement of the protein leading to the protection of the DNA from the damaging effects of the energy pulse and hence the single cell "Prokaryote" species was compatible with its environment.

As the environment underwent further changes the operation was repeated with extension of the DNA molecule as further genes and RM's were formed with multiple specimen variations occurring as they were subjected to changed environmental exposure effects. As gene change occurred the next variation, of necessity, supported the previous causing the DNA and its evolving specimen and hence multiple specimens (the species) to survive (live) in a particular environment. (The established genes were passed on from one generation to the next as the nucleotide of the existing half of the gene was strained beyond its elastic limit and on the other half being replaced when the vibrating energy pulse was applied the inherent weakness of the original nucleotide led to the rapid weakening of the new half, reinstating the gene. Once installed in a species DNA a gene is permanent, whether functional or not).

The RM and gene reacts to an energy pulse over and above that of the normal incoming heat energy and where a gene is activated in a cell/differentiated cell the invigorated reaction results in the formation of a clump of protein around which the relevant gene etc. is entwined (Histone) plus overflow into the chromatin. This clump of protein thrusts the gene clear of the chromatin protein making it available for RNA replication (Attach. 4).

The result is the observed differences in the chromatin is erroneously thought to code for characteristics as does the DNA whereas it is actually muffling the incoming environmental energy pulses with frequencies relevant to the differentiated cell, protecting the DNA from extensive damage and therefore leading to survival of the species.

Due to the writhing motion caused by the energy pulses the extended DNA molecule ends eventually joined forming the DNA circular format of most of the prokaryote species now in existence.

This is the basic process of all formats of life, obscured in certain species as they have evolved into more complex forms of the various Eukaryotic species as their environments have developed. The basic principle is in all cases the DNA reacts to energy pulses that are directly attributable to its environment and it has been evolved to resist the damaging effects of these energy pulses and as persistent slight environment change occurs the DNA reacts and is evolved making it and the specimen compatible with the environment and thus survival of the DNA. (As environment effects are delivered chaotically and the DNA are extremely sensitive to change the conscious life characteristic is a reaction to chaotically varying changes and is present in varying degrees depending on the evolutionary development of a species as its DNA reacts to its environment).

Some two billion years ago a prolific single cell species had evolved that had the reaction of producing sugar via the sun excitation/chlorophyll process e.g. lichen etc. (The sugar was/is a conduit for environmental heat energy). In the presence of this process a cell evolved with the capacity to produce energy by chemically reducing sugar. During this period other single cell bacteria like species evolved that directly consumed sugar from the single cell lichen species and indirectly released sugar molecules. With the input of energy via the sugar there was an output from these cells in the form of propionic acid and this acid was an indication of the presence of and an inciter for the evolving species to absorb sugar when in the presence of this output.

This is the Mitochondria cell species and eventually it invaded a host prokaryotic cell (symbiotic relationship) with the energy pulses it released activating the DNA of the host cell leading to a more vibrant and active reaction from the cell over the full 24 hr. period. Eventually a further similar single cell invaded this infected cell and with its increased energy utilization the invaded cell was capable of supporting the DNA of the secondary cell as well as its own. This was the process involved in the establishment of the combination "Eukaryotic" cell and the activating process for this complex cell was the forerunner of the sexual act.

During this period of rapid change the eukaryotic cell evolved sophisticated mechanisms including the protein processing ribosomes, organelles etc., with the nucleus and the cytoplasm being effectively reliant on control from the DNA as it reacts to the environment influence. Additionally the circular DNA molecule underwent strain to such an extent enclosed in the initial cell casing (the nucleus) it ruptured into various lengths forming chromosomes encased in the protective chromatin protein.

If a mutative effect occurs caused by radiation, virus infection, asbestos etc. to the DNA nucleotide pairs that is beyond its renewal capacity the protective protein output formulae is disrupted (Chromatin, as the DNA alphabet or code for the production of protein is changed) and incoming environmental energy pulses can then penetrate to the damaged DNA resulting in further rupture of the nucleotide pairs and rapid fire propagation of the cells with damaged DNA and dysfunctional performance in duplication of cells as they attempted to build parasitical organs (Cancer).

(From point A, the beginning of life phenomenon to point B, the present, it seems some forms of cells and their outputs are extremely complicated and unexplainable, however with an understanding of the basic process and an appreciation of incremental changes over the four billion years time span the mystery unfolds)).

Gradual increasing exposure to different conditions (environments) various cells underwent processes of adaptation (evolution) as the relevant energy pulses drew reactions and development of genes from the two sets of DNA resulting in their survival with individual output s controlled by genes being compatible. A major requirement for survival was the exposure to sustenance (sugar) to support its energy requirements and various methods dependent on circumstances developed e.g. plants extract requirements from the soil and air whilst being energized by the sun excitation/chlorophyll process and forerunners of fish species drifted to their sustenance obviating to a large degree the need for movement control, and direction to for acquisition of sustenance therefore evolution of extensive logic and brain power.

Other Eukaryotic cells were evolving along different processes depending on environmental exposure and this included a species that eventually evolved into the mammals.

For this developing species exposed to damage from the increasing environmental effects the outputs from the cells of necessity meant an increase in energy necessary for survival. Exposed to an environment where it was subjected to the propionic acid isomer molecules from the bacteria that incited the recognition of the presence of and absorption of the sugar the increased energy availability resulted in the multi cell species surviving. As the species environment further developed it gradually evolved, increasing the numbers of multiple cells reacting to increasingly different energy pulses delivered from primary and developing internal chemical means (secondary) in support of the preceding cells (differentiation) and becoming more complex and during this developmental stage the supporting bacteria was incorporated into its makeup resulting in mammal species dependent on a gut flora symbiotic relationship.

This symbiotic relationship persisted and does so today aiding in the recognition and the mental outputs to acquire sustenance with its propionic acid molecules output. Without this output, during the conception stage to the embryonic stage when the cells acquire appropriate receptors there would be no access to the cells and therefore to sustenance. When the umbilical cord attaches to the fetus the mother supplies the variety of chemicals required for the developing differentiated cells until the time of birth, where outside environmental influences take over its development until the child reaches maturity. With this system the mother's health and practices can influences the child's development over its lifetime (e.g. if mother is placed on a diet in first semester or suffers from lack of sustenance her chemical reaction, passed on by chemical means can result in child developing cells that instigate behavior resulting in compulsory eating and therefore obesity. Without this symbiotic relationship mammals would not exist, however the advantages outweighed the negatives as the process resulted in an efficient access to sugar that supported their development.

As the evolving species requirements to access survival sustenance increased the senses along with limbs were evolved in support of mobility with cells being established with the capacity to build flesh, bones etc. allowing access to sugar and hence survival. Further it necessitated the characteristic of logic and reasoning to acquire this sustenance and cells evolved in support of an existing communication system between the cells (Eventually lead neuronal cells of the brain). (The DNA of each and every cell and therefore the cell reacts selectively with an output due to external energy pulses (the input) in a mechanistic manner as physically induced reactive distortions of the DNA nucleotide pairs of various descriptions are installed (The RM and genes) that are recognized and interpreted leading to reactivation of the cells by ongoing inputs and therefore outputs resulting in survival of the species. As the environmental input changes the output capacity requirement results in DNA molecules dedicated to a task and additional cells in support of the previous cell when an increasingly complex physical output for survival is required.

When a mental (conscious and unconscious) response is required capacity to store information in an accessible mechanistic format (memory) enabling a logical (accessing and sorting) reaction is required to enable survival the capacity is provided by extension of the memory (RM) portion of the DNA molecule (Junk) and an increase in number of supporting "Glia" cells. Some of these evolved in support of and dedicated to various tasks such as the senses (consciousness) whereas others evolved dedicated to subconscious mental outputs such as balance control in walking. As the environment affecting each sense increased becoming more complex additional supporting cells were evolved that had the capacity to react to the increasing information and relay it selectively in funnel like fashion to the relevant neuronal cell containing the relevant memory, if any. Inputs from the various senses were coordinated as they evolved, leading back sequentially to the cell with the initial output source, capable of causing a physical reaction.

Repeated persistent exposure to a subject caused mechanistic distortions of the DNA nucleotide pairs representing a recognizable input and depending on the degree of exposure and distortion these strained nucleotide pairs recover at a variable rate. With restricted exposure they eventually recover completely, but with repeated persistent exposure they become virtually permanently distorted i.e. long term memory and if the total inputs are between these limits graduated memory terms occur. If the incoming information alerts a memory and it requires a response an action output is initiated. If an action is seen as spontaneous such as suckling it is due to the specimen's ancestors having been subjected to the associated stress repeatedly and the inherited junk nucleotide pairs permanently distorted. Additionally to convey these inputs to the memory centres, synaptic connections between the glial cells are injected with a protein capable of conducting the appropriate signals but stifling others (Attach. 6). If this protein is not reinforced during further controlled inputs allowing quick reflex actions, it is destroyed and the components recirculated by maintenance cells. With the onset of ageing these maintenance cells may deteriorate and there is an uncontrolled buildup of the protein chocking and killing of cells resulting in Dementia i.e. long term memories still available to a degree but short term badly affected.

This line of species in time diverged as different groups were exposed to slight persistent environment changes trending to evolve them into numerous species as they that became divergent. Due to the necessity to support this change it lead to the adaptation of further characteristics, resulting in different sustenance acquisition demands and consequently mental control and physical abilities of various complexities. Mankind owes its existence and mental capacity to the complex environments over the course of history it has been exposed to.

This accounts for many of the earlier evolved genes having similar outputs but genes with variations as the species diverged. It also leads to the inescapable conclusion that all characteristics including mental ability and capacity are installed by a mechanistic physical process as a result of exposure to the ongoing nurturing environment of a species and nature is a result of the selectively accessed recorded effects of the reactions by the DNA to the specimen's nurturing over generations. (Mendel's Laws of Inheritance). Therefore mental reactions cannot be entirely controlled by an individual.

Further conclusions

1. Cells do not and cannot spontaneously propagate. (Law of Physics, Cause and effect)
2. Specimens with parents from divergent environments may inherit incompatible genes resulting in serious genetic problems e.g. Diabetes type 1. MS. (Attach. 7).
3. Exposure to persistent energy pulses outside of those normally encountered in a species environment leads to extensive genetic damage with resultant problems such as cancer, because Evolution has not evolved protein with arrangement of amino acids molecules to protect against the man made energy pulses.
4. Species do not adapt to their environment, their DNA molecules are activated and react to the environmental energy effects, otherwise they would not be in existence.
5. Darwin's Theory of Evolution that evolution occurs by natural selection does not hold up in that it means, it is a spontaneous process and by the laws of physics every effect must have a cause.
6. Stem cells do not exist. The effect is achieved when cells with common DNA molecules are activated, during their pluripotent stage as they propagate, by the correct application of the applicable energy pulse required to activate the pertinent effects.
7. Clostridia bacteria, part of the gut flora emits Propionic acid molecules as its output that ingress brain cells in their role of assisting in identifying sustenance and aiding in its acquisition. This bacteria is physically passed on from mother to child and takes three years in the child to become stabilized. The acid molecules. Carbon based, are isometric i.e. may have different energy content depending on their structure as a result of the state of the evolution of the bacteria. Bacteria as evidenced by recent developments are prone to rapid evolution and antibiotics when administered during their stabilization period is part of their environment and this results in the output production of a changed evolved isometric molecule. When chemically reduced in the brain cells they release energy signals that are different and confusing resulting in dysfunctional physical and mental

reactions (Autism). Drastic change of sustenance (diet) to that normally experienced by a migrant mother can also induce the problem in a young child as the bacteria has evolved to deal with the mothers initial diet.

Attachments

1. "Introduction Wave Genetics, Genes, DNA" etc. Extract from http://www.soaceandmotion.com/evolution-biology-wave-genetics.htm
2. Extract from "science of Zoology" Mc Graw-Hill Book Co. Penetration of Protein molecule by energy pulses.
3. "Quantum Tunneling" Extract from http://www.newscientist.com/article/dn-electrifying-claims-for-dna-are-dashed.html
4. "Histones" Scientific American Dec 2003.
5. a. Low intensity microwave radiation as modulator etc. by Vojisavljevic V. Pirogova E. Cosic I.
   b. Demodulation in tissue, the relevant parameters and the implications for limiting exposure. By J. Silny.
   c. DNA fragmentation in human fibroblasts under extremely low frequency electromagnetic field exposure by Focke F. Scherman D. Kuster N. Schar P.
   (Papers demonstrating the phenomenon of non environmental energy pulses bypassing protein and damaging DNA).
6. Extract from "Biology Today" Random house publishing Extract is "The Molecules of Life" (Protein).
7. Evolutionary divergences resulting from Environmental Exposure.

Purpose and substance of each enclosure

1. As a result of experimentation demonstrates the DNA is reactive to environmental energy impulses

2. Demonstrates the layout of the amino acid molecules of the complex protein molecule and their symmetry with format of pulsating energy and what it means in respect of cancer.
3. Establishes how the generation of electricity is confined to a restricted length of the DNA resulting in a controlled signaling or switching of the gene
4. Clearing of the gene from the clutter allows efficient production of the RNA.
5. Experimental results establishing that contrary to authoritative views that LF radiations such as emitted from smart meters do not represent a danger to human populations, they do.
6. Indicates that Protein in any of its many forms is utilized by the cells to undertake functions leading to its and hence its species survival.
7. Demonstrates just how sensitive DNA is to exposure to environmental habitats.

ALZHEIMERS, DEMENTIA

◇◇

Initially at the beginning of the life process DNA molecules formed in the barren landscape when an enzyme assisted chemical process resulted in readily available elements i.e. Hydrogen, Nitrogen, Oxygen, Carbon and Phosphorous forming into molecules of RNA. Eventually two RNA molecules linked together via hydrogen bonds, becoming a DNA molecule as a series of their compatible nucleotides A to T and G to C became available. The nucleotide arrangement was susceptible to resonating in the vibrating effects of the incoming pulsing energy representative of the environment and eventually suffered damage. Over the course of myriads of years as a chemically induced reaction of the DNA molecule suffering damage a cell formed resulting in partial protection for it against the damaging effects of the incoming energy pulses that constituted its environment. During the course of this development atoms of Potassium were incorporated into its cytoplasm. With the availability of potassium this situation applies to every cell that has ever appeared.

The presence of the Potassium and Carbon result in a magnetic field (from the always present radioactive atoms K40 and C14 in these two elements) supplementing the existing magnetic field.

The DNA molecules within the cell have become progressively adapted by a hit and miss process resulting in survival as various means of physical protection and mental reactions evolved. Outputs from the cells provided these protective and mental reactions against the potential damaging effects of the incoming environmental energy pulses leading to survival of the DNA molecules. As the DNA molecules, are exposed to the incoming energy pulses of a range of frequencies that they have evolved to withstand and are normally encountered in a species environment, output reactions leading to survival are initiated from the cell as the pulses access the now appropriate genes via associated nucleotide pairs (The Regulatory Memory RM). (When multi cell eukaryotic species appeared specific environmental energy pulses resulting in particular differentiated cells were delivered, by the enzyme assisted chemical reduction of particular chemical molecules within the cell that were actually conduits for the energy of various environmental effects e.g. sugar etc. This resulted with the input of controlled energy pulses establishing further genes that initially controlled the reaction of the cell with the production of an output leading to its survival and this output continues to be released as the repeating environment effects of pulsing energy continue to afflict the gene resulting in the output of the cell contributing to the survival of the species as it becomes commensurate with this environment effect).

The format of the DNA molecule is such the ladder like rungs (Nucleotide pairs), extremely sensitive to environmental pulsing energy are subject to vibrating as a reaction to the resonating effects of the incoming energy pulses. (Vibration and resonating is the most common physical phenomenon in the universe).

The controlled vibration of specific nucleotide pairs (rungs) caused by the incoming fluctuating pulsing energy results in an electric current being generated in them as they resonate in the magnetic field. The nucleotide pairs conducting this electricity are surrounded by an induced force field. This force field flows and ebbs, directly mimicking the effects of the environment. The electric current is constrained to specific nucleotide pairs that have been evolved, tuned to the frequency of the particular incoming pulsing energy, by a materials property known as quantum tunnelling. The force field surrounding these specific Regulatory Memory (RM) nucleotide pairs is associated with a gene and this force field switches on, or draws a reaction, from the gene and the effect is the gene, which is a record (Generational memory) of the potentially damaging capabilities of the incoming environmental pulsing energy effects initiates an output from the cell's cytoplasm that results in the necessary protection or reaction against this particular aspects of the environment and hence survival results.

The function of a cell is one where the basic process is always the same i.e. incoming environmental pulsing energy causes reactions from the DNA molecule that initiate outputs from the cell that result in its survival. (A cell does not nor can it function spontaneously i. e. every effect has a cause, basic law of physics)

Despite the apparent complexities of the evolved detailed operations of a cell, all the various differentiated cells in any species have evolved with integrated outputs that are always involved in the survival of the DNA molecules and therefore the species. (This is a hit and miss process that continues until the involved nucleotide pairs become established in a fixed arrangement (coded) when the incoming, persistent environmental energy pulses are now reflected in the cells survival output (Evolution)).

In the beginning of the development of eukaryotic species when multiple cell species appeared it resulted in a communication system between the cells coming into existence with coordination of their outputs. As the numbers of communicating cells gradually evolved and increased several events were occurring and in the case of the forerunners of mammals the following is a depiction of these events.

1. With the appearance of multi cells and the incorporation of the mitochondria into them with its energy releasing capacity, sugar was being produced by developing plant life Algae and Lichen. The availability of this sugar was influenced by a single cell bacterium attacking these plants and as a result sugar molecules were released to which this developing multi cell species was exposed. As a waste product from the bacterium, molecules of propionic acid were also being released.

2. The propionic molecule, being extremely small penetrated into the cell via the opening in the membrane created by the second cell as it combined with the initial cell. Being energy rich the acid molecule was chemically reduced by the mitochondria resulting in the cells reacting to the pulsing energy and the initial receptor in the membrane enlarging, allowing the ingress of sugar molecules, thus increasing the available energy levels over a 24 hour period. With unrestricted energy now available the cells were energised and began to multiply

3. As the eukaryotic species slowly evolved with cell numbers increasing the requirement for the developing specie's cells that were shielded from direct contact with the environment to acquire increasing supplies of energy from external sustenance

arose. For this to occur an ability for movement to access sustenance (Source of energy) developed and it resulted in controllable appendages along with the senses slowly evolving to facilitate movement within the species environment. The process here was that incoming energy pulses reflected from the sustenance slowly evolved the senses involving the establishment of the appropriate genes and the relevant RM aspects of the DNA molecules.

4. The initial multiple cells that were being evolved into the control and communication characteristic, developed receptors where messages (or the effects of environmental energy influences) were passed on. These receptors were the synapses and facilitated the communication system between what eventually became the controlling (brain) cells. As the environment became more complex increasing numbers of evolved cells generated reactions initiated by incoming energy effects from the various senses evolving, resulting in a protein being injected into the synaptic gap allowing the message effect to be transmitted (conducted) from cell to cell in a downloading, sequential and integrating effect to specific cells (Neurons)(The original communicating cells). These neuronal cells became integrated into the physical control systems of the appendages that were evolving.

5. Direct control of the limbs and flexible components was necessary to achieve the response required (Survival) and as the complete picture was loaded into the neuronal cell (Attachment 1.) the characteristic of logic arose, which is an analysis of many installed effects of energy pulses (Recall or memory) and this results in the initiation of action and thus survival via controlled energy (Electrical) pulses that created reactions in the controlling cells of the evolving muscles. This is mental awareness associated with the inputs to the sense

6. Further facilitating the ingress of the sugar molecule into the initial cells a cell type evolved that produced a protein, insulin. As the initial controlling batch of cells were evolving, (potential brain cells), these support cells producing the insulin were closely associated with them and do so today as a individual type of differentiated cell in the makeup of the brain. As the species developed further reactive cells involved with the

physical aspects of survival, the pancreas organ, evolved with cells capable of an output of insulin to facilitate the functioning of all the physical supporting organ cells.

7. The characteristic of memory is due to the property of the nucleotide pairs that are tuned to specific frequencies, being distorted by the vibrating effect of the incoming energy pulses that are commensurate with an effect of the environment. The property of the nucleotide material is such that the distortion effect affecting the nucleotide rungs results in them recovering slowly (Recovery rate) and the more the available incoming energy pulses are distorting them the longer it takes to recover. These distortions are then "recognised" as the incoming associated environmental effect related to the senses (Short term, medium term and long term memory) accesses them

8. As the environment of a species became more complex incoming energy pulses via the senses resulted in more "memory" capacity of a mental output control capacity evolving resulting in survival of the species. The result was the installation of further RM nucleotide pairs to the DNA chromosomes, by a chemical substitution replacement and addition to, of what were now the in tune nucleotide pairs (The "junk" DNA) to the incoming energy pulses, however these energy pulses are not persistent enough to result in the installation of numerous genes and therefore a limited range of physical outputs resulted. This Junk DNA is utilised in the Glia cells of the brain and they have the function of downloading the gathered information in a consolidating formation to the appropriate neuronal cells. The more complex the species environment the greater the numbers of evolved glia cells result and increased capacity of the Junk DNA.(The junk DNA does not and cannot retain all the memory effects installed in it prior to the reproduction process taking place, however its format may result in behavioural traits being inherited. Further the process is such it leads to an ability to recall memories and when those memory effects are such they have been installed from unacceptable events they result in repeated recall with permanent imprints being installed in the glia cells extended memory nucleotide pairs as the person affected tries to come to terms with it. This is PTSD that can

result in serious mental reactions and the condition is inheritable by following generations).

9. The process is such the various brain cells associated with the different senses react to incoming energy pulses of a variety of forms as conversion processes have evolved to convert them into a uniform format and they are gradually integrated together as an overall picture into the neuronal cells that may result in a long term memory when exposure to the environmental effect is continually repeated.

10. The normal exposure to an immense variety of incoming environmental effects, such as a travelling car, a charging lion, trees in a forest and myriad every day events, result in sufficient information for recognition being installed in the neuronal cells to elicit an output resulting in survival in this environment, however these imprints are insufficient to result in permanent memory of the details of these event, however sufficient is retained to recognize these events without in depth detail when they are repeated. In fact if this did not occur, it would result in a complete overwhelming of the brain cells and the ability of the species to survive would disappear.

11. Subjected to infinite numbers of chaotic incoming environmental effects to the senses the protein injected into the cell synapses normally becomes surplus and as the cell is once again activated, by inputs that are not often repeated, the protein previously produced as part of the procedure is thrust aside by the newly manufactured protein. To overcome this problem in normal circumstances, supporting brain cells have evolved that have an output that breaks down this protein resulting in its makeup materials being recycled.

12. With repeated exposures the conducting protein in the synapse tends to become consolidated forming permanent pathways to the relevant neuronal cell, leading to rapid memory recall of familiar people, objects etc.

13. As age encroaches the various cells fall off in efficient operation as mistakes in the nucleotide pairs are copied etc. and the cells become tired (Ageing).

14. The result is the glia cells drop off in their capacity to clear the manufactured protein as it is produced; clogging the cells (Tau

tangles), the support cells have an efficiency drop off and the functioning cells are overcome by the build-up of protein. This is loss of short term memory with associated malfunctioning of the brain. (Alzheimers, Dementia)

15. The incoming environmental energy effects can no longer penetrate and download to the neuronal cells and the short term memory is seriously disrupted, whereas the embedded long term memory system retains some semblance of functioning.

How to deal with the problem?

As the problem is caused through the drop-off of the efficiency of the cells, boosting factors that are involved in their efficiency could involve increasing the potassium levels getting through to the brain and therefore increasing the magnetic field available for the efficient operation of the cell and secondly Insulin dosage could possibly create a situation where the intake of sugar from the blood flow could provide extra energy for the more efficient operation of the cells. The potassium levels could be increased by the consumption levels of potassium rich sustenance such as Bananas and Tumeric spice.

The question should be asked, is Insulin a treatment for diabetics alone or should its operating characteristics be used to improve the functioning of cells thereby increasing their efficiency, making its use a valuable adjunct for many treatments.

ALCOHOLISM

Consumption of any substance over a long and persistent time basis causes a reaction from the cells of a species. In the event it is a chemical that can be broken down in the cells with an energy release it is recognized by any species as a sustenance intake that it must evolve to, as part of its environment, so that it can survive. For the complete process of evolution refer to.

1. http://evolgenmutsuic.com
2. Theory of Life and Evolution

Alcohol has the same materials (elements) in the molecules as does sugar i.e. C, H and O. It is also bound together by positive energy, as is sugar

Given a user consumes it frequently enough and in sufficient quantities, the mitochondria DNA of the cells, responsible for extracting the energy from sugar, by way of producing the necessary enzymes to break down these molecules is gradually evolved becoming adapted to its use as a substitute for the sugar, and this occurs because the released energy pulsations are of a different frequency and the nucleotide pairs involved begin to evolve to enable the DNA to cope with this new environmental affect.

This process occurs as the enzymes normally used to break down the sugars and release the energy are applied to the molecules of alcohol, a changed signatory pulsation of energy occurs, that is responsible for applying changing strain to the nucleotides of the mitochondria DNA, thereby eventually causing evolution of the mitochondria in the offspring. Further, when normal energy pulsations are persistently disrupted the changed pulsations take effect in the evolutionary process by accessing the chromosome(s) and the DNA is influenced to produce a system requiring alcohol as an energy source as all species tend to adapt to their intake The pancreatic insulin production cells are also affected and the DNA nucleotides of the chromosomes responsible for the production of insulin will evolve in the offspring, aligned to alcoholism. Other secondary processes similar to this may affect bodily functions (As all cells throughout the specimen use energy) and be associated with the tendency towards inherited alcoholism

When the above ancestry applies, where both parents have been abusers or carry the effects in their DNA (Genes etc) as mutations, the offspring can inherit the insulin problem according to Mendel's laws of inheritance. As the mother is involved in passing on the mitochondria the direct effects of alcoholism are inherited from the mother i.e. it is a partial maternal- linked problem

This is evolution in action and the person is developing to be a user of alcohol instead of the normal energy source, sugar, and the offspring to inherit alcoholism when the mother in particular is an alcoholic. Exposure of the offspring to an alcoholic environment probably encourages the young to drink (The result of nurture becomes nature), additionally where the young have a mother with inherited alcoholism, stressful events can lead to alcoholism appearing. It is not necessary for the alcoholism to be apparent in the mother if it has been controlled; however it can be passed down through generations (Alcoholism, like all other characteristics can be inherited, depending on the amount of exposure, both to the individual and the ancestral generations).

There are several consequences of this such as

1   Mental capabilities are energy driven and these capabilities are disrupted because the mind cells involved do not function efficiently
2   Alcoholism is present for the person's lifetime.
3   Control of alcoholism is extremely difficult as it disrupts the person's willpower by the bodies systems interpreting the need for alcohol as being necessary for survival.
4   As an equivalent amount of alcohol carries more hydrogen atoms and less of oxygen atoms than sugar, the hydrogen draws

the oxygen creating a dearth of oxygen within the cells and producing more water which is excreted, thus causing a drying out of the body and a requirement for more liquid intake which can lead to heavier drinking. Instead of the waste gas created within the cells being carbon dioxide it is carbon monoxide or a like poison product.

5   Alcoholism may not be seen as present very early in life to some other genetic characteristics, but may be activated by an initial exposure to alcohol.

6   Although the mitochondria is not subject to extensive evolutionary change, there is sufficient difference, caused by regional varying sugar molecules to trace the originating area of the mothers cultural ancestry and hence the individuals ancestry.

7   Drugs such as cocaine, ice and amphetamines are constituted of C, O, H and N elements and are conveyed in the blood stream to the cells where they have a similar effect by disrupting their functions. Provided the parent has indulged sufficiently prior to having offspring the children can inherit addictions

Carcinogenetic effect

When the ethanol alcohol molecules are reduced by an enzyme, byproducts of acetaldehyde molecules result.

A further enzyme then reduces these molecules to acetic acid under normal circumstances, however if the supply of the enzyme is reduced by ageing etc. or there is insufficient enzyme being produced to cope with the amount of acetaldehyde due to excessive drinking, the molecule(s) may penetrate to the DNA molecule and in some cases disrupt the gene(s) responsible for initiating the protective protein (chromatin) that is found wrapped around the chromosomes in the cells nuclei.

The coding (codon) of the gene representing the required formulation for the protein is then disrupted and the proteins muffling capacity protecting the DNA molecule from the incoming environmental energy pulses is reduced further resulting in diminished protection, increasing cyclically, with the result being an increasing rate of propagation of dysfunctional cells i.e. cancer. (Refer to paper Penetration of Protein molecules by Energy Pulses (and cause of cancer)).

AUTISM, ITS CAUSE

After viewing a documentary featuring ideas on they're being a faulty gut microbiological condition association with the brain resulting in Autism I realized the hypothesis I have been working on answered the questions posed as to the cause of the problem. The hypothesis establishes the basis of life and how it evolves and from this the causes of problems such as cancer, diabetes, dementia and behavioural and mental problems can be resolved. On learning of the manner in which Autism strikes the patients and of the gastroenteritis problems associated with it I took a closer look and came to the conclusion that although my hypothesis was still valid I had overlooked the fact that the bacteria flora associated with the human species (and other mammals) were species and therefore had ignored the fact that they also were subject to evolution on change of environment.

On studying the documentary and associated website articles I noted that there seems to be a prevailing consensus that environmental conditions and effects are disassociated with genetics. This could not be further from the truth as the DNA molecule reacts to environmental energy and this involves all species and the response from the molecule is the setting up of a defensive mechanism when genes etc.are evolved that help combat individual damaging effects of the incoming environmental energy of different frequencies and amplitudes. This is in fact why and how all species are compatible with their particular environment. Species do not adapt to their environment, they react to it via the DNA and are driven by it and to conclude they do adapt to it is completely illogical as there would be no life in the first place.

Note

Physics phenomena and engineering principles facilitate the establishment of this hypothesis. Normally outside of the ken of the biological and genetic professions the engineering profession can vouch for the role they could play in the process of life and evolution, as can I as a retired professional engineer.

The attached hypothesis explains the cause of Autism

Life is due to the ongoing reaction of DNA molecules to the damaging effect of the energy pulses of its environment (All effects of a species environment access the DNA molecules by way of energy pulses, from either primary or secondary chemical means). The life process is the continuous reaction of the extremely sensitive molecules of DNA as the potentially damaging environment energy effects chaotically occur causing changes of the energy pulses involved resulting in rapid reactions from the DNA that guides a response from the involved cell to counter these effects. (This applies to both physical and mental characteristics). The response from the cell is an unchanging procedure involving the DNA process with evolved variations i.e. production of specific protein molecules, hormones, enzymes etc that result in the protection and survival of the DNA and hence the species. To this end the complex protein molecule produced, where required has the capacity to muffle the relevant energy pulses thereby containing the damage to the DNA of the specific cell resulting in an extension of the life of the cell before it propagates. In instances where specific chemicals enter the differentiated cells by way of a receptor the chemicals are reduced releasing energy pulses that are muffled by the protein chromatin resulting in controlled fatigue of the nucleotide pairs and hence extended life of the cell. Where the chemicals released from the sustenance during the digestive process act as a controlled conduit for the environmental energy output then the chemicals are reduced in the cell. Energy pulses received via the senses are subtle additions to the original environmental energy sources and hence the lifetime endurance of brain cells. An integrated output from the evolved differentiated cells as a result of the various established environmental energy inputs results in the controlled life process of the specimen. A disruption to this process causes the process to become dysfunctional.

Evolution occurs when there is a persistent minimal variation to the specie's and hence the DNA's environment and therefore the incoming environmental energy. It proceeds when the DNA's defences (Protein etc, a protective shielding) are unable to immediately counter, muffle, this change to the energy pulses, as the specimen does not have protein evolved for this purpose and damage occurs to the DNA's nucleotide pairs most closely associated with the varied incoming energy pulses. An enzyme assisted repair response occurs to the DNA section involved (Regulatory Memory (RM) and Gene(s)) resulting in a change and additions to the DNA nucleotide pairs and its property of reacting by initiating production of adapted protein etc from the cell that is once again protective thus allowing the DNA's survival in this developing environment i.e. the DNA extends with developments to the genes and RM nucleotide pairs) and this results in variations to the specimens characteristics, with when multiple specimens are exposed over time and reproduce, evolution of the species. (Evolution does not entirely occur by natural selection. In some cases it is responsible for the survival of the species).

Countering this damage from the environment is always achieved in the same manner with the incoming energy pulses activating a signal via the RM nucleotide pairs that gets an appropriate reaction when the gene(s) are activated to initiate a response from the cell leading to survival. (A gene is the damage effect on the nucleotide pairs of the DNA by the environmental energy responsible for its development and then it represents a guide to the formulation of the protein etc that protects it from further damage or aids in the species survival when exposed to an energy input from one of the five "senses" sources) In the physical sense this reaction results in proteins, hormones, enzymes etc that constitutes the species with its coordinated cells that have evolved with DNA genes that initiate a response when activated by the appropriate incoming switching signal as a result of the relevant environmental energy pulses inducing an energy field from the RM nucleotide pairs (See Theory of Life and Evolution). The proteins etc that are produced in this manner accommodate designated tasks both physical and mental that lead to the specimen's survival and hence it's DNA and reflects the species evolutionary history. This then accounts for all of the various species evolving as they are exposed to variations of environments.

The following facts have to be borne in mind when considering this hypothesis

1. Species do not adapt to their environment. They are the result of their DNA reacting to the effects of their environment. (Otherwise life would not exist in the first place. It is also the reason all species are compatible with their environment).
2. A cell cannot and does not spontaneously propagate. (Law of physics, "Every effect must have a cause")
3. A cell cannot and does not react spontaneously causing protein etc to be produced (As above)
4. Every living species has DNA and is therefore subject to the same principles
5. DNA is the single common factor (Albeit in different arrangements) and is therefore the "living" entity that is an activated molecule responsible for all life as it reacts to the environmental energy, resulting in its and the species survival.

Now to "Autism"

A human is not a single entity but is an individual relying on the coordinated effects of bacteria for survival as they; the bacteria species' rely on the host. This is necessary for mammal life to continue as the process has evolved over thousands of generations i.e. they are symbiotic species

A human rapidly migrating, along with its bacteria flora, to another country is subjected to a change of environment resulting in many hidden problems of which Autism is one. A change of diet is part of the change of this environment as the sustenance has generally been raised in this environment and is therefore, amongst its roles, a conduit for the particular environmental energies involved in its growth and development. As the chemicals resulting from the digestion of the sustenance are chemically reduced in a cell it releases controlled energy pulses commensurate with the environment it was raised in. Over sustained generations the human cells DNA molecules have evolved to be compatible with this diet.

Humans are not subject to rapid evolutionary change of environment as are bacteria, however the bacteria "Clostridia", from the evidence presented, having evolved in the gut to a particular diet emits chemical molecules of Propionic acid that is commensurate with this environment, on change of the humans diet it rapidly evolves with the resultant chemical output emitted now a toxin or incompatible to the host (The Propionic Acid molecules are Isometric i.e. they have the same chemical makeup, but the constituent atoms are tied together with different energy bonds). The original chemical was developed as the clostridia evolved in the bowel cells was exposed to the sustenance and therefore the bacteria cells output was representative of this particular aspect of the original environment. As the human evolved certain brain cells were kick-started and became activated by the propionic acid chemical molecule released by the clostridia bacteria and as a result of them being activated by the incoming sustenance which was representative of the acceptable diet when the chemical molecule was reduced in the cell the released energy impulses established reactions in the "Junk" DNA molecules as the embryo developed the brain organ, of memories that identified suitable foodstuffs and controlled behavioural traits. (It may also be involved in establishing the need for sustenance (Hunger) and control of the eating function). As the basic characteristics of foodstuff acquisition on which the species depends involve behavioural, physical and mental responses e.g. hunting etc all these associated responses evolved to be implicated. (This aspect of Environment change on the evidence presented is not as decisive as the Anti Biotic change).

Given rapid migration of humans a drastic change of sustenance may occur with an offspring's gut flora susceptible to change as it develops in the first three years of its life. The child's genetic makeup however is representative of the maternal parent's original environment and hence the propionic acid Isomer molecules are toxic, resulting in distorted reactions when the molecule is reduced in the cell, releasing dysfunctional energy pulses.

Installation of memories and action responses, depend on the incoming environment influences effecting the initiating brain cells, but as established in the paper the process is always the same i.e.

the RM response switches on the gene(s) leading to survival and in the case of glial brain cells when the "Junk" part of the DNA chromosome (Evolved extended RM) is activated the response includes the gene(s) guiding production of a protein by the cell to be injected into the synaptic connections suitable for transmitting the relevant energy pulses to the specific neuronal cells for action and recognition purposes.

Given the incoming chemicals from the clostridia are no longer compatible (the genetic makeup of the patient has not evolved) when they are reduced in the brain cells they are involved in the release of energy pulses that initiate a dysfunctional switching signal from the RM nucleotide pairs to the gene(s) that results in the production of a defective protein by the cell response that is impaired to such an extent energy pulses, involving behaviour, sustenance recognition and physical responses are garbled, eliminated or added to via the synapse and sent on. This is Autism.

This is the dietary aspect of the problem, however it is well recognized that bacteria quickly evolve to survive in an anti-bacterial agents and as the clostridia bacteria is extremely resistant to these agents then it is rapidly evolving to this change of environment and instead of producing a functioning chemical a "Toxin" is produced with similar results as above. Further the patient being afflicted by both causes can exacerbate the problem.

Suggested remedies

1. Revert to diet of cultural country and remove evolved clostridia bacteria and replace with bacteria from an unaffected specimen from that culture after treating the patient to counter the emerging clostridia from the spores. (Hopefully before the evolved clostridia are mature enough to propagate).
2. Stop excessive use of Anti Biotic agents.
3. Examine the possibility of cultivating clostridia capable of producing a normal brain activating chemical compatible with the *patient's* genetic makeup and injecting it on a regular basis to the patient. Kill off existing evolved clostridia and attempt to overwhelm the germinating clostridia with normal clostridia.

1. "Wave Genetics: On the Wave Structure of DNA and Resonant Interactions of Genes and Environment" Extract involving Prof Pjotr Garjajev's (Russian Biologist) and team of researchers work on DNA molecule having resonant characteristics.

2. Extracts from "The Science of Zoology" 2nd edition McGraw Hill

3. "Environment Exposure Divergence" various extracts.

PENETRATION OF PROTEIN MOLECULE BY ENERGY PULSES. (AND CAUSE OF CANCER)

The only reactive entities involved in every living species are the DNA molecules and as such they are the life giving factor and Proteins, hormones etc. are the building blocks of life.

All normal environment effects a species is exposed to activate the DNA molecules resulting in output reactions from its cells. These effects are all, on critical examination, by way of primary energy pulses emanating from the environment, and/or secondary chemical energy conduits e.g. a sugar molecule when chemically reduced within a cell releases controlled energy pulses synonymous with its evolved species characteristics and the environment it was established in and grew in. These chemicals are in fact, as well as supplying the various chemicals required by internal cells, conduits for the environmental energy pulse effects normally experienced by the species.

The energy pulses involved are from sources including various types of carbon based chemical molecules obtained from sustenance and those chemicals developed within a eukaryotic specimen's cells as a result of a reaction to energy source inputs including primary environmental energy. Cells evolved in a complex process utilizing available elements from the species environment by the DNA molecules reaction to these sources of energy and the outputs are relevant to the species survival in its environment.

The forerunner of the origination of life and genes occurred when on the initial formation of the chemical DNA molecule in the primeval swamp, on exposure to resonating pulsing environmental energy (heat) certain of the nucleotide pairs of the molecule became physically damaged by the resulting vibration effects and an enzyme aided chemical reaction occurred continually replacing these damaged nucleotide pairs, with pairs of a different orientation (Regulatory Memory, RM), until they were in harmony with the frequency and amplitude of the incoming environmental energy pulses whilst adjacent pairs although not physically destroyed were strained beyond their elastic limit and this was the initial gene. Carbon C14 radioactive isotopes and Potassium radioactive isotopes K40 were also present in the swamp environment, providing a magnetic field.

The incoming energy pulses continued to vibrate the RM nucleotide pairs that were now tuned to resonate harmonically through this magnetic field generating an electric current in them that was restricted to the involved nucleotide pairs by a materials property known as quantum tunneling.

Accompanying the electric current a generated force field surrounding the conducting nucleotide pairs was rising and falling in cohesion with the environment effects.

This generated force field highlighted the damaged state of the strained (gene) nucleotides "Switching" on a chemical reaction (mRNA), that supported their survival and hence the DNA molecule in the environment. This response was the initiation of a chemical reaction, resulting in a basic protein (The Chromatin) that eventually protected the DNA molecule by a hit and miss process with its arrangement of integral amino acid molecules that protected the DNA molecule from further excessive damage from the incoming environmental energy pulse effects.

The life process began when the environment began to slowly change and became more complex resulting in incoming pulsing energy of various wave lengths and amplitudes, relevant to the environment but able to penetrate the initial protective protein, chromatin, as it had not been evolved to provide the necessary protection, but as the environment changes were controlled and persistent, retrievable damage was caused to the nucleotide pairs activating an enzyme assisted chemical repair and replacement process thereby establishing an increasing number of them involved in gene development of the DNA molecule resulting in an adapted output of a controlled process of protein production etc. activated by them, that protected the DNA molecule from further destructive damage by these pulsing energy inputs resulting in its and the evolving species survival in this environment. This included reactive genes involved in establishing a complex arrangement of the protein chromatin, thereby providing protection against an ever increasing range of environmental energy pulses as the environment became more complex.

Eventually multi cell species came into being and with the mitochondria process providing energy 24 hours a day the Eukaryotic species took off with the phenomenon of gene installation increasing.

Additional genes and their accompanying RM's were added and adjusted, extending the DNA molecule as damage repair took place when an enzyme assisted chemical replacement process took place resulting in changed and additional numbers of installed nucleotide pairs, in a hit and miss process that eventually ceased when equilibrium occurred due to the protection provided by the now adapted and increasingly complex original specifically formulated chromatin protein that surrounded the DNA chromosomes in the cell nucleus i.e. the incoming energy pulses could no longer gain significant access to the DNA nucleotides and the process of adjusting coded nucleotide pairs temporarily ceased. (This is a continuing ongoing process, evolution and life). The gene nucleotide pairs of the DNA molecule (chromosome) are now in an arrangement representative of the damaging vibrations (Genes are the "coded" arrangement of nucleotide pairs that have been strained beyond their elastic limit by the vibration effect of the incoming relevant environmental energy pulses, and are now representative of the incoming energy pulses due to the hit and miss elimination process that results in the adaption of the gene's nucleotide pairs to this coded arrangement resulting in the cell's production of appropriate proteins etc. The protein produced has an arrangement of the constituent amino acid molecules that provides a muffling function resulting in survival of the species in its environment. The gene has, through the ongoing development of the process become representative of the effects of these energy pulses and therefore is a record/ memory of the effects that caused the damage). With the ongoing repetition of these energy pulses the arrangement of nucleotide pairs (RM,) associated with the gene, not strained beyond their elastic limit but 'evolved to be tuned to the frequency of this incoming energy pulse are activated and "switch" on the relevant gene by the electricity generated process and the gene is initiated to activate an output process from the cell of protein etc., that results in the physical damaging effects of this energy pulse input on the nucleotide pairs being nullified or a mental reaction that is protective or leads to survival of the species, and this output then results in the cells extended survival in the environmental conditions.

Incoming, controlled environmental energy pulses result in the DNA chromosome genes of various differentiated cell types being activated and initiating a controlled output (Both mental, where applicable, and physical) from related organs etc. that in every Eukaryotic species represents its characteristic of life output resulting in it being compatible with its environment. Any slight persistent change to the energy pulse input from a persistent environment change results in this energy pulse penetrating the chromatin protein protection, causing further adaptation to a gene(s) and the RM nucleotide pairs by a hit and miss process that eventually results in the changed energy pulse no longer causing an immediate damaging effect as the newly adapted coding resulting in a changed formulation of the cells protein, eventually resulting in the species being compatible with this environment (evolution) as the effected cells outputs are adjusted. (If the change is drastic then the DNA molecules are damaged beyond control and the species suffers extinction).

The illustration page 60 The Science of Zoology (McGraw-Hill book company) is representative of a protein molecule and in this case it is the Chromatin in the cell's nucleus that is produced, encasing and protecting the cell's DNA chromosomes from the excessive vibration damage potential that the environmental energy pulses have. The controlled energy pulses then access the DNA molecules and activate the relevant genes, achieving controlled survival reactions (life).

Referring to the article, the diagrammatic line intersecting the "R" factors is representative of the helix trajectory of an evolved arrangement of the amino acid molecules that in themselves are three dimensional. Pre generation (Inherited) genes and incoming environmental energy pulses involved in the evolution and the functioning of the DNA genes of the differentiated cell are involved in the chromatin protein coding formulae and hence the production of this protective arrangement of the amino acid molecules making up the protein molecule. The arrangement of these molecules provides a damping effect for the incoming normal environmental related energy pulses by sequentially absorbing energy from them and hence reducing the energy that finally impacts the DNA nucleotides thereby extending its physical life. When a persistent small change to the existing environment energy pulse occurs the RM and gene coding responsible for protein production is adjusted by a hit and miss event as the energy pulse is not protected against and penetrates to the DNA chromosome resulting in controllable damage and an adjustment to the DNA's chromosome gene nucleotide pairs coding until chromatin protein protective molecules, of a slightly adjusted amino acid molecule arrangement result with the protective capacity being reinstated. (This energy change results in evolution as the cells overall output is also adjusted as gene change or addition occurs).

Note

The chromatin for humans may be a conglomerate of proteins providing protection against the specific energy inputs associated with the genes of the differentiated cells performance and the common heat energy input in any differentiated cell.

As the protective protein (The chromatin) has been evolved to dampen existing energy pulses the application of significantly different energy pulses to those normally present in the environment e.g. man induced energy pulses including radiations outside of those normally experienced in an environment, are not protected against and can penetrate the protein, they may cause irretrievable damage to the evolved arrangement of the nucleotide pairs coded for replacing the chromatin protein coating that surrounds the chromosomes, and this coding then spells out the formulation of a protein molecule that no longer provides the full damping effect required against the impact of the normal environmental energy pulse on the chromosome's nucleotide pairs.

The effects of these energy pulses as they gain further access creates damage to the coded nucleotide pairs resulting in the chromatin protein formulae deteriorating further with increasing access for the energy pulses as the process goes into cyclical mode. Over the propagation of generations of this cell the increasing irretrievably destructive damage to the nucleotide pairs of the chromosome results in expansion of the problem to associated chromosomes with their eventual break up (Ref) into numerous minor chromosomes as the cell is rapidly driven to uncontrolled propagation by the increased energy effects accessing the genes that activate cell production. Due to the now stressed remaining nucleotide pairs suffering an increased rate of fatigue due to the energy impact by passing the now non damping effect of the chromatin protein and subsequent division of the DNA molecules, it results in the time of the life span between the initial development and eventual propagation of the dysfunctional cells being drastically reduced.

Initially the protein output of the cell is affected and not matched to the immune cells sensory expectations of the cells output from the established DNA chromosomes and being irregular the affected cells are normally dispensed with, however occasionally an individual damaged cell may be missed resulting in the immune system being overwhelmed as the propagation of this cell results in numbers of increasingly defective cells rapidly propagating, with hugely increased energy demands as they rapidly build.

These cells are now running amok, overwhelming the immune system and tending towards building a conglomerate of parasitical organs.

Depending on the extent of the damage initially incurred it may take many generations of cells for this process to be fully established, hence cancer may not occur until many years after the causal event.

The cause of the problem besides radiation may be stable chemicals (e.g. Asbestos, nicotine etc.) intruding into the activating gene's coded arrangement of nucleotide pairs for chromatin resulting in a dysfunctional formulation of the protein with an arrangement of amino acid molecules that no longer adequately muffle the incoming energy pulses, Viruses when present in the DNA molecule may disrupt the nucleotide pairs coding and intervene with the resulting production of chromatin protein being dysfunctional. An inherited mutation may also cause the problem and these may appear later in life as the differentiated cell is brought into operation.

Ref: "Chromosomal Chaos and Cancer". by Peter Duesberg. Scientific
          American. May 2007

Suggested Treatment for cancer

Olive Oil is a lucrative source of Mono-unsaturated fat that when the oil is present in the diet is converted into macromolecules (Chylomicrons) in the lower intestine, that are a large combined molecule consisting of a combination of fat and protein that are then dispersed in the blood stream. Anti-oxidants are also present in the oil.

A cancer cell utilizes up to fifteen to twenty times the energy that a normal cell does as it rapidly propagates and divides.

Supply of this energy is provided to the cell by sugar molecules along with primary energy and these molecules are conducted to them by the blood flow.

The sugar molecules are chemically broken down in the cell releasing energy pulses in a controlled process that supplies, along with primary heat energy input, the required energy for functional processes but the differentiated cell is activated via its relevant DNA genes when evolved incoming carbon based chemical molecules are chemically reduced releasing energy impulses of a particular frequency that activate the Regulatory Memory RM nucleotide pairs resulting in activation via the associated gene of an output from the cell providing physical protection and therefore survival of the species in this aspect of the environment. This process occurs in conjunction with the activation of other differentiated cells when other relevant evolved incoming chemicals activate complementary outputs from the cells that on activation of the appropriate genes an integrated process results causing reactions (mRNAs) that guide the different cell's complex outputs resulting in survival of the DNA molecules and hence aids in the survival of the species.

If the blood transporting the sugar around the body has Chylomicrons (Macromolecules) suspended in it they are also transported around the body. They are composed of fat and protein produced in a complex process involving the mono-unsaturated fat provided by the olive oil and this eventually results in, if in sufficient quantities, the affected cell being dealt with in a quasi-similar fashion as the autoimmune system deals with infected cells. The protein component of the chylomicron macromolecules lock into the cell receptors and the fat components inundate the cancerous energy hungry cell preventing it being accessed with the required sugar resulting in it being deprived of its excessive energy requirement and in its eventual demise.

Additionally the anti-oxidant destroys oxidants thus reducing their influence in cancer occurrence.

Unlike in the case where cancer cells that are treated with radiation or chemo therapy result in problematic side effects and stray effected cells being missed with cancers reappearing and spreading, every affected cell in these conditions self-inflicts its own demise.

Appropriate dosage and period required are not known and are subject to experimental verification, however it appears 9 table spoons a day for an 85 Kilo person can be tolerated but it is suggested that the maximum dose possible would be the most effective.

Notes:

1. Support for this theory comes from the fact that Mediterranean populations whose diet encompasses a lot more olive oil than other western societies and statistically suffer fewer cancers.
2. It is possible that the macromolecules cannot penetrate the brain screen barrier due to their size and therefore it may require brain specialists to investigate the possibility of injecting them.(later established they can).

DIABETES

<><><><><><><><><><><><><><><><><><><><><><><><><><><><><><><><><><><><><><>

ITS CAUSES

To be read in conjunction with:
   2.   Theory of Life and Evolution.

Diabetes' are problems arising from the malfunctioning of the energy provisioning system of the species cells that may lead to inheritable problems and eventually problematic evolution.

There are three recognized forms of diabetes. These are:

A. Type 1 or Insulin dependant diabetes (Previously known as Juvenile diabetes).
B. Type 2 or mature age diabetes.
C. LADA diabetes or Latent Autoimmune Diabetes of Adults

Diabetes' are the ongoing failures of the energy (sugar) processing system of the cells. They are the result of the process breakdown involved with the glucose intakes of the cells and the resultant energy production. Through farming, mankind has inadvertently introduced two directly opposed constituent's i.e. excess fat and excess sugar into his diet. This situation has resulted in diabetes type 2 that may lead to offspring inheriting the problem and eventually causing evolutionary problems.

Secondly, worldwide there are ten basic evolved variants of the sugar molecule, the local one of which the indigenous populations have evolved to be compatible with. Through rapid migration, crossbreeding of specimens from different environmental habitats occur and their inherited genes, responsible for the pancreatic output of insulin, may not reflect the active autoimmune system comparable genes evolved in the partners or fore bearers habitat and thus the insulin is incompatible leading to an attack on the cells producing it (The Auto immune system now senses the pancreatic cells are infected) Problems can also occur over generations as individuals persistently consume sustenance from different overseas environments causing an evolutionary reaction.

Autoimmune systems have been evolved to protect a species against debilitating influences such as foreign outputs from the cells induced by bacteria and viruses; however as with any system problems can occur. Type 1 and LADA diabetes is due to the autoimmune system identifying the individual's insulin being produced defectively by a foreign entity and consequently destruction of the cells of the pancreas that are apparently malfunctioning occurs. An indication distinguishing LADA from type 2 is that sufferers are thin indicating a possibility the problem may be associated with the cells of the digestive system (liver) with the cells responding by allowing excess sugar into the blood stream due to being influenced to evolve by persistently high sugar levels in the blood over generations causing production of superoxides resulting in conditions leading to diabetes. LADA is possibly programmed differently to type1, occurring later in the life cycle.

The life process is about the survival of DNA in its particular environment. DNA in some format is the constant reactive entity throughout every species. The geometrical features, except the length of the DNA strand and the arrangement of the nucleotide pairs, which are variable, are always similar, being adapted to react to the normal varying energy pulsations encountered in their environment. With the secondary environment like condition that have evolved in the cells of eukaryotic species, due to the presence of the mitochondria (DNA) and its ability to release energy from sugar via an enzyme process with approx.130 pulses per molecule (Varying from glucose molecule type to molecule) as it is broken down providing for measured release of the energy via the ATP molecules means these energy pulses access the appropriate nuclei DNA (Pancreatic and Reproductive) by the differentiation process and vibrate specific nucleotide pairs (Refer to "Theory of Life and Evolution") that results in an output of the protein, Insulin from these cells. The incoming energy levels in the cells activates a controlled reaction in the pancreatic cells resulting in insulin production, which then facilitates the access of controlled levels of glucose to all cells. The controlled role of the insulin in allowing glucose to access the cells cytoplasm, via receptors in the cells membrane, where it is chemically reduced and releasing the required energy may be significantly affected by fat obstruction. To maintain control of the glucose necessary for energy consumption within the required levels the liver cells have also been evolved to control the release level of glucose into the blood stream.

Countering the potential damage the incoming pulsating environmental energy can inflict on the potentially resonating molecular DNA, a process involving reactions from the DNA including outputs from relevant differentiated cells has evolved. These reactions are due to the chaotic fluctuating energy effects that access the DNA. The fluctuating reactions are the life effects where various portions (genes etc) of the DNA are tuned to react to specific energy inputs that have fixed definitions i.e. frequencies etc. When the environment is slowly and persistently changing, leading to different conditions (Changed or additional energy frequencies etc) the DNA is damaged and rebuilds incorporating new genes etc as a reaction to these changed energy pulses with the relevant nucleotide pairs once again, becoming tuned to the now persistent frequencies, and on the continuing application of these energy producing coordinated inputs, physical or mental outputs are induced. The physical outputs consist of protein of many and various descriptions and the molecular arrangement of their molecules is tied to the energy exposure effects on the genes initiating their production. The cells outputs of protein have an amino acid molecular arrangement specific to the individual having been established in response to the wave energy effects recorded in the genes of the involved individual specimen of the species having been passed down from its ancestors. The DNA genes were and are a result of and reflect the damage the environmental energy pulses it has been exposed to have caused, and these genes are now a template for the formulation of the protein necessary for survival in its environment. Production quantity is controlled by the genes relevant RM as it reacts to the varying environmental energy influence it is being subjected to (Primary or secondary). The proteins produced then aid and assist the individual to resist further damage from this prevailing environment energy with its fixed frequency, by adapting the specimen and eventually the involved specimens (the species), resulting in its survival in this environment. This is Evolution of the physical characteristics of the species when individuals have been exposed to prevailing similar environments and subsequently reproduced. As the gene etc is added or adapted to further change of the prevailing environmental energy the resultant output of protein molecule is also adapted. The habitat the species occupies is the source of much of the prevailing energy effects it is subjected to and therefore its physical characteristics are controlled to a large degree by the habitat.

Further to this the energy inputs involved with mental abilities have caused reactions from the DNA developing cells with mental output capacities and cells with supporting output functions i.e. proteins such as serotonin, dopamine, synapse gap filling protein, etc facilitating these functions

A DNA reaction to the pulsing energy release from the molecules of the incoming glucose is part of the survival process of the species in its environment, supplying heat energy (secondary system for eukaryote species) when they access the cells cytoplasm with the mitochondria DNA reacting with an ATP enzyme aided reaction of chemically reducing this glucose. With the chemical reduction of the glucose, with approx 130 pulses of energy per molecule released, the DNA responds throughout the specimen by becoming activated and vibrating in the released heat energy pulses.

The energy release process from the sugar is common throughout the cells however the Pancreatic and liver cells have evolved with differentiated functions with genes of their DNA reacting to these energy pulses and producing a protein (Insulin) with the property of controlling the finite energy output of the cells. The cells function when activated by the input of the prevailing environmental energy that initiates the supply (switches on) guiding and controlling the appropriate amount of protein required i.e. insulin in the pancreatic cells case. When the normal energy pulses are interfered with, due to the supply of the glucose to these cells being reduced by fat deposits clogging the receptors the energy impulse inputs are reduced and the output of the cell drops and if the condition causing the problem persists (A changed environment, secondary) the DNA evolves genes controlling the supply of a defective quantity of Insulin.

Normally in the case of the Pancreas organ, sufficient supplies of the protein Insulin is produced, rendering an adequate supply of glucose to each cell relevant to the survival of the specimen in its environment. A persistent variation or change to the glucose supply as it enters the cytoplasm draws a reaction as a change to the environment responsible for the energy supply (via the glucose) and is coped with by the DNA being adapted to counter the perceived damaging effects of the resultant energy pulse changes (In the case of excess fat in the diet, a reduction of energy pulses due to the drop in available glucose occurs). The immediate effect of the fat restricting the ingress of the glucose and the inherited problem of loss of capacity to process the glucose due to reduced insulin is type 2 diabetes. In some creatures, where glucose is short in their diet they have evolved to be languid with a lack of energy. In the remainder of the cells throughout the body the gene(s) involved in the production of insulin in the pancreas are neutralized by the system of gene differentiation.

The DNA of the male reproductive organ and the female eggs (gametes), are normally similarly accessed over the lifetime of the parent specimen however when excess fat plays a part persistent reduction of the energy pulsations from the chemical secondary means (sugar) accesses the reproductive cells DNA, adding or adapting nucleotides pairs of the appropriate genes when failure to elicit harmonic responses eventually results in damage to the appropriate nucleotide pairs and in time a resultant gene and RM correction occurs This damage reaction is due to an environment where less sugar energy is available and this is reacted to as an environment change and the quantity of insulin being supplied is improperly corrected with an adjustment to the DNA genes etc that eventually relate to the pancreatic organs production of insulin. This result in reduced insulin and excessive sugar in the blood stream of the offspring when the diet is not adjusted causing internal damage and reduced energetic responses from the individual. This situation is resulting in an epidemic of the evolving condition of Type 2 Diabetes now spreading to young people.

Note

Type 2. Diabetes condition was originally known as Mature Age Diabetes as it appeared later in life. It can affect older people who have not initially inherited it, but it is becoming more prevalent in the young for several reasons,

1. Poor diet and sedentary lifestyle leading to overweight young and obesity
2. Inherited from young, overweight and affected pre-conception adults (It does not necessarily appear in the individual at a very young age, as inherited characteristics appear in order of evolution and this condition is appearing very late on the human historical scene)

There are ten variations of basic sugar protein molecules throughout the world, with possibly further variations in basic foodstuffs and the people who have been evolving in different environments have become adjusted over many generations to the slight variations applying to them, in these sugars. Given the rapid migration of many of these people from different environments they are not genetically suited to the new environment, including generally to the sustenance, of a variation of sugar molecules. When reproduction occurs between people of different habitat origins their offspring may have pancreas cells with inherited DNA molecules controlling production of a protein, Insulin, evolved to suit an habitat, but that is unacceptable to the inherited autoimmune system resulting in it attacking these cells. Another problem may be the separate DNA of the Mitochondria is not fully suited to its task of reducing the particular glucose that is made available and as a result functioning poorly. This then is Type 1. Diabetes.

This intermixing of DNA traits present in migratory populations can lead to many problems and because of its nature may take many generations to resolve and it is most certain that as long as rapid migration of individuals is continuing then exacerbation of the problem in the population will continue to surface. The problems may persist over generations and because of this reason it may not be initially apparent as to the reason they exist. Type 1. Diabetes is a fault with the process and therefore appears at the first opportunity, whereas Diabetes

Type 2. may be an inheritance problem and therefore appears in order of evolution or as it is established.

A related problem also exists* where an excess of sugar rich food pressures the liver to deliver excess glucose and gain entry to the cells resulting in the mitochondria malfunctioning and producing free radicals (superoxides). These superoxides then restrict the access of the glucose to the cells interior resulting in the excess glucose remaining in the blood stream and doing damage. It appears that this is another process by which Diabetes may evolve due to reduced sugar in the cells, however it may not be a regular cause as it is necessary for the specimen to be overfed on a regular and extensive basis, whereas a common normal western diet contributes to the fat and sugar conditions in an intermittent ongoing basis resulting in inherited Type 2. Diabetes.

The Glial brain cells usually persist throughout life as they have evolved, coping with less strenuous energy inputs. The result is they retain conscious memories (recognizable strain of nucleotide pairs) and react when an initiating input is applied. If they were to be replaced these memories would be lost. To support the functioning of these cells a process of transferring information (Energy pulses) by injecting protein across the synapses gap has evolved. This system occasions the removal and recycling of the protein and facilitating this other type brain cells have evolved As their efficiency drops off, due to ageing (wear and tear) or a similar problem to that above occurs a build up of this excess protein occurs. As mental processes are an energy dependent process, and rely on intermittent production of protein to efficiently transfer information via the synapses, the problem of being able to function arises. (This is a problem apparent to diabetics when low on glucose). As a result of this loss of efficiency, several problems may occur, partially due to the protein produced for the installation of new information clogging and damaging the involved cells. The failure to destroy it results in malfunctioning and death of some cells, and they become unable to install new information, although having been installed previously in downstream cells (neuronal) with repeated inputs a permanent pathway has been established resulting in earlier repeated information being available. This breakdown of cell efficiency is dementia etc. and has been observed in some instances as being related to obesity which can, as previously stated, be involved in effectively blocking incoming glucose that normally provides the necessary energy and therefore hindering the brain cells in their role of processing and passing on information.

Notes

1. Type 1 diabetes is not apparent immediately in babies as they are initially programmed to survive on a diet of milk albeit human milk Inherited characteristics can appear as programmed, not just at birth.
2. Domestic animals, whose diets have been adjusted by man, are also showing similar problems.
3. Type 2 diabetes may also occur if the available glucose is in low supply or facilitating chemicals are in short supply over a sustained period. This problem may occur in the womb or through out a specimen's life.

PROPOSED TREATMENT

For early cases of type 2 they may be treated by dietary means, however for inherited cases the following applies: The incoming environmental heat energy stimulus, provided by the incoming glucose, activates the process of the appropriate gene(s) resulting in the functioning and output of the Pancreas cells to produce insulin, that enables the glucose to initiate the output of most of the cells and consequently of the body, when it accesses the mitochondria DNA is exposed to the glucose molecules enabling the cell to produce the required enzymes dedicated to breaking them down with the release of further energy required for the cells processes.

In the early stages of development of the young the randomly selected pluripotent cell initiates the establishment of the pancreatic organ cells and they have the same DNA and gene specification as all other normal cells throughout the species, and have differentially evolved to produce the required quantity of the protein insulin as a reaction to incoming energy pulses, this normally results in enabling controlled glucose consumption and therefore the energy production needed from the body's cells. If the required glucose access to the cells is restricted by fat, the dedicated DNA (nucleotides) of the pancreatic cells reactions can be for the DNA to evolve over generations to produce genes that produce reduced supplies of insulin. Production levels of cell outputs resulting from energy based reactions is reduced to match the perceived stimulus, which is insufficient for the available glucose, and consequently the body cells fail to access sufficient glucose causing hyperglycemia problems while the unused glucose in the bloodstream causes hypoglycemia.

An examination of the DNA of a type 2 diabetes sufferer identifying the gene now reacting to the energy impulses from the glucose in the area (regulatory memory of the genes) established as being responsible for the quantity of insulin production in comparison to a non- sufferer will indicate an evolved difference in the nucleotide pairs. Splicing corrected nucleotides into the patient's pancreas cells that are about to replicate and injecting them back into the sufferer's pancreas will see them build replacement cells capable of producing increased quantities of insulin without having extreme rejection issues. Diabetes type 2 is an ongoing problem of diet that can be inherited as well as developing during the individual's lifetime.

In the case of diabetes type 1 the pancreatic cells will have the functioning gene nucleotide pairs defining the insulin formulae (Insulin is a protein and is therefore never the exact same molecule as for any other specimen), that are incompatible with the autoimmune system, as they have been evolved in different environmental circumstances. Each autoimmune system of the parents should be genetically compared with the patients and if possible the sufferer's pancreas cells adjusted by cell gene therapy to most closely resemble the nucleotide pairs of the DNA controlling the insulin of the parent with the autoimmune system that most closely resembles the patients.

L.A.D.A. type diabetes will require further examination to definitively establish where exactly the problem is in the cells before establishing a treatment process

Notes

1. The possibility of tackling the problem by this method is demonstrated by the procedure of producing medical supplies of Insulin, whereby the DNA of bacteria has been modified to produce human type insulin.

2. Unfortunately some evolution of the mitochondria DNA is also taking place and this requires further investigation to establish whether it is a problem and as to how it is handled.

3. Inherited alcoholism, and drug addictions that maybe caused by similar processes, are also involved.
   (This is why the Aboriginal people of Australia (And other Indigenous people) are so involved)

4. Splicing nucleotides into the responsive portion of DNA may eventually be superseded by the system of adjustment by an energy corrective method. (Russian research authored by Prof Pjotr Garjajev).

5. Due to the process of Evolution described above there is a basic human genome but it is never exactly the same from individual to individual. For this reason establishing whether a mutation is present by a method of comparison must proceed with caution.

6. Many complications have developed as the species has evolved but the basic system is where the DNA molecules have reacted to the environmental energy pulses resulting in evolution of the species and therefore survival.

Refer to   www.spaceandmotion.com/Charles-Darwin-Theory-Evolution.htm
Then to "Evolutionary Biology, Wave Generation."

Reference:

*   Garvan Institute paper
"The Free Radical that triggers Insulin Resistance and Type 2 Diabetes"
(Proceedings of the National Academy of Science Sept 2009 Dr Kyle Hoehn, Prof
David James and team)
"Theory of Life and Evolution" (unpublished) by John A. LeRoy

THE CAUSE OF MS

‹‹‹

Genetics and the environment are NOT two separate aspects individually affecting life. They are both synonymous with the process of life as the DNA molecule physically reacts ensuring its survival in ITS environment, as all aspects of any environment are by way of radiant energy pulses, resulting in species with all of their resultant characteristics. Such diseases as multiple sclerosis diabetes, dementia, cancer etc, in fact all genetic problems affecting living species are the result of disruptions to the process with the species becoming more complex as evolution driven by the changing effects of an environment occurs.

Starting at the beginning of the life process, when the complex molecule of DNA came into being as a result of chemical processes its ladder like format resulted in its nucleotide pairs being prone to vibrating as a reaction to the incoming environmental radiant energy pulses of the normally present heat and light. It was in the presence of Potassium a commonly occurring element. Potassium always has an isotope of radioactive (K40) in its makeup (Most cells of every living species has K or in a very small % a substitute i.e. Uranium). Additionally isotopes of radioactive Carbon (C14) were present and this conspired to provide a magnetic field available to each cell The vibration effect resulting from the environment energy effect resulting caused resonance of the nucleotide pairs of the DNA through this field resulted in an electric current being generated with a surrounding energy field that provided a signaling effect commensurate with the radiant energy pulses and hence the environment. This switching effect highlights a damaged length of nucleotide pairs, the arrangement of which is developed as a result of a sustained change of environment and hence changed radiant energy pulses, destroying nucleotide pairs that were originally harmonically resonating to the original fixed frequency energy (Ref 1). An enzyme chemical aided reaction then rebuilt and added nucleotide pairs in a hit and miss process until the molecule with the involved pairs once again becoming harmonically responsive to the incoming effects of the environment. The involved pairs (The Regulatory Memory, RM) having extended and moved along have left behind nucleotide pairs mechanically extended beyond their elastic limit (The gene) due to their previously extended exposure to the vibrating radiant energy pulses. The gene nucleotide pairs then have changed electrical conductivity properties and are representative of this damaging effect of the ongoing environment. The switching or signaling effect from the RM then results in the recognition of the gene and the formation of a chemical molecule (RNA) that initiates a response (proteins etc) that occurs in the cytoplasm of the cell via the evolved ribosomes. The responses, based on the damaging effect of the environmental energy, have characteristisc that ensures the continuing survival of the DNA in the environment. (The resulting protein etc molecules have evolved with a chemical makeup resulting in the survival of the DNA in the damaging effect of the various aspects of the environment the specimen and its ancestors have been and are being exposed to). This situation still occurs today as the propagated DNA molecules of every newly propagated cell (Pluripotent), which are constructed of basic passive elements, are reactivated as they are forming and the existing damaged nucleotides of the original half of the chromosome result in the damage being reinstalled by vibration as without the presence of the chromatin protein the energy pulses freely access the DNA molecule and as a result the energy pulses act on the weakened nucleotide pair, straining them beyond their elastic limit

The extended length of the human DNA the "junk" nucleotides pairs etc constituting 93% of its length have evolved as harmonically responsive to various new energy effects related to the environment that require mental responses resulting in the survival of the species and hence the DNA. The basic process remains the same however as the "junk" DNA in the evolved brain cell is activated utilizes associated genes where production of a protein is initiated for facilitating the passage or downloading of energy impulses through the cells via the harmonically in tune nucleotide pairs when vibrated. The proteins selectively produced guides the input from the involved senses to the appropriate neuronal cells by way of the synaptic connections of the glia cells, the arrangement of which has evolved as an increasingly selective download system, coping with the increasing demand as the environment has developed and extended. The energy pulses generated are in the form of electrical pulses. This process results in

A. When exposure to the environmental effect is spasmodic and rarely experienced the initial strain distortion effect on the relevant nucleotide pairs quickly recovers. Depending on the frequency of a similar exposure with resulting strain the recovery takes longer. This is long-term short-term memory etc.

B. The protein injected into the synapses of the relevant cells if not reinforced by similar exposures deteriorates and is destroyed and recycled by supporting cells. If continually exposed the relevant proteins become consolidated and are available for rapid recognition to take place as the memory is consolidated and the incoming energy pathway defined. (This system is necessary to cope with the enormous amount of incoming, mostly irrelevant information)

Note

With ageing the support cells can deteriorate and a buildup of excess protein and waste surrounding and within the cells normally involved in the installation of new information (memory) process may result in them being damaged and dying whilst the neuronal cells holding sustained memories are not as affected. This is dementia.

When activated the neuronal cells deliver an appropriate electrical signal to the muscles etc. resulting in action and hence survival. The efficient conduct of these signals over the nerve fibres, both outgoing and incoming and within the brain cell connections resulted in the evolution of the Schwann cell that produces a protein with electrical insulating properties. (Ref 2, 3)

The development of cells and eventually multi cell "species" with a number of cell types evolved (differentiated) to cope with the variety of the incoming energy inputs resulted in the DNA survival. These species diverged, incorporating various materials and being adapted as a reaction to the varied energy changes as they were exposed to varying environmental affects i.e. evolution.

Species don't adapt and evolve to suit their environment; the environmental energy effects of their exposure evolve them via their DNA, resulting in their survival and the system of evolution is as described above. (If this were not so there would be no life)

The rapid, ongoing varying reactions of the DNA to the ever-fluctuating (chaotic) environmental energy effects are the life phenomenon.

With the advent of eukaryotic species, where the development of multi cell species took off, a coordinating communications system between the cells came into being. With the advent of an environment change resulting in the necessity for a species to become mobile and thus access sustenance, a capability to apparently logically think and control this mobility (the brain) evolved, driven by the energy input from the environment, as it was reflected from the sustenance, and a communications system, controlling the evolving cells of the fins, arms legs etc by interfacing between them and the brain evolved by development of the existing system. (The reaction and adaptation to any environment of a species is entirely dependent on the interaction between the DNA nucleotides and the pulsing energies of various fixed frequencies of its environment. The DNA is the only physical reactive entity to environmental affects in all-living species). The communication system is now known as the Neuronal nerve system and it has gradually been evolved over millions of years originally as a complex system conducting energy signals to facilitate the control of mobility and thus survival.

As the evolution of the eukaryotic species commenced the multi cells were being evolved, all requiring an output to support the development and survival of the species and thus it's molecular DNA in the damaging effects of the increasingly complex environmental energy.

Resulting from its exposure to the direct energy of the environment was the evolution of the outer cells (skin) to produce an output of protective chemical 7 dehydrocholestrol to counter the direct incoming environmental UV B energy in the frequency range of 270-300 nm. Further reaction between this chemical and the ongoing energy produced a chemical molecule now known as Vitamin D. The quantity of this vitamin can vary significantly depending on the skin cells exposure to different intensities of this environmental energy.

Processed by the species metabolism the pro-active form of this vitamin supplies a triggering pulsating energy effects to the internal cells, where necessary, when they are no longer directly exposed to the environment. This occurs when the vitamin accesses the cell's cytoplasm via the established receptors in the cells outer membrane, where it is chemically reduced, releasing energy pulses equivalent to the initiating environmental UV B energy involved in its production. These energy pulses then initiate an output from the cell via the controlling response of its DNA. Over a sustained period of exposure to a variation of the energy pulses the DNA nucleotide pairs (Regulatory Memory and Gene) evolve, reducing the damaging effects. This adaptation is also handed down to offspring via the inheritance process, eventually resulting in the ongoing evolution of a species.

As one of the initial processes established in the eukaryotic species the process evolved, with the vitamin D combating the damaging effects of the UV B energy.

Evolution of most species takes place very slowly and is a controlled process relevant to their environment that should apply to both parents so that the genes etc are compatible. With the relatively rapid migrations of humans from different environments to a new environment disruption can occur to this system when interbreeding between parents with genes evolved in significant different environments takes place and the "Schwann" cells produce the insulating protein that may not be compatible with relevant inherited genes of the other parents autoimmune system evolved to be compatible with a different performance of the vitamin D effected Schwann cell (Other parent's and/ or ancestors environmental exposure). The immune system cells react to the protein output from these cells as it is not compatible resulting in a threat being sensed of dysfunctional cells and a chemical (protein) response destroys them.

The process in detail is the White Blood Cells (T) part of the Auto Immune System encounters the dysfunctional protein (recognized as such by the fact the protein molecule produced by the Schwann cell does not match with the normally dormant genes for that protein associated with the auto immune system DNA. (Evolved protein commensurate with the environment) The T cells as they are then replicating and during this process have a pluripotent stage develop with a receptor of the outer skin of the cell capable of accommodating this "dysfunctional" protein. (The cell reproduction system involved with differentiating a cell, normally confused as the "Stem Cell" system). The energy pulses from this protein molecule when released in the cytoplasm of the autoimmune cells initiate the production of a left-handed version of the "dysfunctional" protein molecule from this cell. The released molecule then locks into the right hand version and this combined molecule shuts the Schwann cell down by smothering it and preventing the ingress of energy releasing chemicals resulting in its demise. At this time the production of T cells is accelerated with their production stepped up. This is a similar situation to the cell being invaded by a virus and the resulting protein output disrupted.

The problem resulting from mismatched genes with the insulating capacity of the Schwann cell protein disappearing, allowing leakage of signals and therefore loss of control happens later in the life cycle as in the life process events appear in the order of their evolution and this problem did not occur until rapid migration patterns appeared.

Inuit people and Laplanders are not as prone to these problems as they have slowly migrated to different environments with sufficient time elapsing for these processes to uniformerly evolve across the population to the conditions. (This also accounts for the more stable Asian, Chinese etc, populations exposed to long-term ongoing environments not being as subject to the problem as populations consisting of basically migrants from warmer environments moving to cooler climates and interbreeding with inhabitants who have evolved to suit the colder climate e.g. Scotland).

Females are more prone to the problem, making it appear as a sex-linked problem of the "X" chromosome. The probability is that as their sex chromosome pair consists of two "X" chromosomes (Male sex chromosome pairs consist of an "X" and "Y" chromosome) one from each parent, the possibility exists for a doubling of the chance for conflict between the autoimmune cells and the Schwann cells. This increased statistically chance of a mismatch between an environment evolved auto immune system and the inherited Schwann cell resulting from a different environment affecting the vitamin D output results in the female's cell output being detected more often as being dysfunctional. The "Schwann" cells then are likely to be destroyed by the immune system, bringing with it the resultant MS.

This situation then indicates, as others problems do, that rapid migration has, on a technical, scientific basis the possibility of severe problems for the offspring of mixed migrant families and these potential problems can range across the whole gamut of human characteristics.

Notes

I am forwarding my papers by post, including "Diabetes, the causes" where it appears as if Diabetes1 is caused via a similar situation.

P.S. If it were a viral problem, males and females would have similar numbers in the populations, and if cooler environments meant the perceived prevalent survival of the virus, then Laplanders and the Inuit people would be more prone to the problem, not less.

P.P.S. Mixed migrant families are not necessarily those from apparently widely differing cultures but may consist of the likes of English /Scottish descendants where historically the possibility exists of Danish/ Roman/ Spanish/Celts etc providing the divergent inherited genes leading to these problems.

Suggested possible treatment

Examine the protein molecule from the Schwann cell of the patient producing the insulating coating of the nerve fibres, and of both parents. After establishing which parent has the most divergent protein molecule to that of the patient, isolate the genes initiating the production of this protein and using a process similar to that used for production of Human Insulin produce sufficient to inject the patient and confuse the auto immune cells into not attacking the Schwann cells.

References

1. "Protein, and its Role in Disease". By John A. LeRoy Unpublished.
2. "Theory of Life and Evolution" by John A. LeRoy Unpublished.
3. "The Other Half of the Brain" by R. Douglas Fields "Scientific American" April 2004.
4. "The Hidden Brain" by R. Douglas Fields "Scientific American Mind" May/June 2011

PHYSICAL DANGERS POSED BY SMART METERS, WIND TURBINES, MOBILE PHONES ETC

◇◇

The cane sugar glucose molecule when chemically reduced in a cell releases approx. 140 energy bursts consisting of radiated energy of fixed frequency. The result is these energy radiation beams impact on the DNA nucleotide pairs of the RM (Regulatory Memory) associated with a particular gene(s)

The process of depending on the controlled reduction of sugar molecules evolved initially as the species developed from being dependent on external energy sources to a secondary source as the cells became partially inaccessible to the external source (Environmental heat etc). This controlled release of energy results in energy bursts at approx. 140 pulses per molecule and this reduces the impact on the DNA of the total energy produced by the molecule. This radiated energy supplemented by various additional energy radiations resulting from the chemical reduction of further chemicals produced as a result of direct exposure to energy radiations from the divergent environment cause the RM nucleotide pairs to vibrate harmonically, and pulsate through the magnetic fields associated with the radioactive potassium and carbon, always present in a living cell, and this results in the generation of electrical energy, confined to the RM and hence operative for the particular associated gene(s). (There are approx 8500/sec atom disintegrations in the normal human specimen creating a magnetic field). These electrical currents induce a fluctuating energy field surrounding the nucleotide pairs that results in a switching or highlighting effect on the gene(s) nucleotides that initiates an output from the cell that is commensurate with the incoming environmental energy (Both by primary and secondary means) and hence these aspect of the environment. There is however 10 basic sugars throughout the world environment e.g. cane sugar and beet sugar etc and on chemical reduction within the cells these sugar glucoses release slightly different energy radiations that are commensurate with their environment. The sugar is then acting as a conduit for the environmental energy effects of heat to the cells, in particular the internal cells of the specimen.

As Eukaryotic species evolved they were adapted to a source of constantly available energy that could be utilized internally i.e. the sugar and as a consequence organs evolved (eventually within the brain and the pancreas) with cells differentiated to produce a protein, insulin, that controlled the ingress of the sugar, the basic source of energy to the various differentiated cells of the specimen. As the developing human species slowly spread, evolving to different varieties, the molecules of sugar they consumed were different, with the resultant changed energy frequencies destroying and damaging the relevant nucleotide pairs of the RM (brain support and pancreas cells) that were exposed and no longer harmonically responding. The destroyed pairs of nucleotides were replaced and added to by an enzyme assisted chemical reaction resulting in an adjusted gene and adjusted RM that was once again harmonically responsive to the energy radiation released by the sugar molecules of the species environment.

The resulting protein output of the pancreas and certain brain cells, insulin, then had a molecular structure adapted to aid in the absorption by the cells of the particular sugar molecule thus ensuring survival of the DNA and hence the species in this environment.

Similarly all of the evolving functions of the cells activated by reactions of the DNA to the changing environmental energy pulses resulted in cells producing protein etc outputs that protected and established survival techniques for the DNA and the specimen

One type of cell evolved that produced protein that was chemically configured to react with any protein that was not compatible with the normal protein producing capacity of the specimens DNA. Being sensitive to when the protein produced by the various differentiated cells was acceptable and therefore were commensurate with its particular environment and would aid in the survival of the DNA in its damaging environmental energy impacts. If the cell produced a foreign protein to that articulated by the DNA genes responsible for this functioning of the (autoimmune) cell it was destroyed e.g. when the protein molecule from a cell is disrupted by a virus then the infected cell producing this protein is indirectly destroyed.

This process of the auto immune system is such it is tuned to be compatible with all of the gene functions that evolve to support each other as they are exposed to a particular environment. It is a slow cohesive process and may be disrupted when due to the ability of humans to rapidly migrate, mating parents adapted (evolved) in different environments produce offspring that may have inherited chromosomes with genes controlling characteristics that are not compatible, i.e. Mendel's Laws of Inheritance come into play.

When the offspring's genes that are responsible for the Insulin have been evolved to be compatible with the autoimmune system sensing capacity then no problem will be present, however if the insulin initiating genes have been inherited that are adapted to the effects of a different environment to that of the autoimmune system the protein (insulin) will be recognized as foreign and the autoimmune cells will destroy the insulin producing pancreas and brain cells, resulting in Diabetes.1.

To combat the problem it appears the most effective method would be to identify which parent's insulin most closely identifies with the patients and then take pancreas cells from the other parent, identify the patients gene(s) responsible for the insulin, remove them and graft the responsible gene(s) from the parent back into the patients DNA and replace them into the patients pancreas where they will reproduce. (There are no "stem" cells; they are the cells that have been differentiated to produce a particular output and on being subjected to the energy radiations i.e. in the case of the pancreas cells that are available from the reduction of the sucrose molecule they undergo the normal life span of the cell and propagate meanwhile producing the required insulin (protein)).

The reverse of this also appears possible by locating the normal source of the propagating autoimmune cells and carrying out a similar operation.

Note

The RM (Regulatory Memory) of a gene(s) is the "Junk" nucleotide pairs associated with a particular gene(s). These nucleotide pairs have evolved to vibrate harmonically with incoming energy pulses from the environment that have fixed frequency and amplitude. As such they respond to the environmental energy effect that has become established in the past and is continuing. Countering (muffling) the continuing damage potential from these energy radiation pulses an output initiated from the DNA is a protein surround of the chromosomes, the epigenetic effect. This effect has been evolved to muffle the normally present incoming energy pulses of the environment. With a persistent change of environment the epigenetic effect is partially by-passed and previously existing vibrating nucleotide pairs are destroyed in a controlled process as a result of them being no longer harmonically responsive and being subjected to unfiltered damaging pulses of the energy radiation and replaced with additional and repaired nucleotide pairs by a hit and miss process until they are now in harmony with the incoming energy pulses and the protein constituting the epigenetic effect is now adjusted by the revised RM of the DNA. This arrangement is then passed on in the inheritance process. As such the nucleotide pairs have been subjected to persistent energy effects to the extent that when the RM nucleotide pairs are readjusted in the evolutionary process they move on with additional nucleotide pairs being added and adjacent pairs that are no longer incorporated into the RM have been strained so they are no longer capable of recovery i.e. strained beyond their elastic limit and these "strained" pairs (The Gene) in the lengths of the DNA then are representative of the damaging effects of the particular environmental energy radiation pulses and for survival of the DNA molecule a protective process, recognizing the potentially damaging effects of this energy takes place. The gene is then a template for the protective process required and proteins etc are initiated and produced in response to the incoming energy of the environment and the output of the associated evolved cell is a protective output as the RM "switches" on the gene, which initiates this output. This then is a "Regulatory Memory" a reaction to an environmental event that has persisted in the past and is continuing as an output from the cell is initiated leading to survival. This output may be virtually a "scab" or a device that ensures the DNA physically survives and hence the specimen in the full effect of the pulsating environmental energy pulses that assails the specimen. This protective procedure over billions of years has developed in small steps that compliment the previous steps, however in sum the impression is of extreme complexity and has resulted in huge numbers of species varying enormously in complexity due to the range of environments they have been exposed to as they have evolved. However it is not set up to protect against manmade and induced energy pulses that suddenly vary from that normally encountered in its environment.

Note

It is the evolved function of the "Epigenetic" protein (Chromatin) to muffle the incoming normal environmental energy pulses thereby constraining damage to the DNA within the cell (It is always about damage control and survival) and disruptions to this process can lead to the impression the "Epigenetic" entity initiates and controls an output from the cell in parallel to the DNA but this is not so, it is a resultant mutation of the DNA nucleotide pairs caused by a radiation energy pulse, or from an effect from outside of the norm bypassing the epigenetic protein that is responsible for a revised output from the cell. The damage (Mutation) of the gene also affects the control of the makeup of the epigenetic protein molecule and as the mutation can be inherited so also can be the changed characteristic output. If the damage from the radiation causes irreversible damage then serious problems may occur

The concept of RM is postulated in several scientific research papers but not understood by the authors as to how or why it functions.

The DNA then always responds in the same manner. The "junk" sections of the DNA react to the energy pulses from the environment (from either primary or secondary sources) "switching" on the gene (Which is a template, representing the damaging potential of the incoming radiant energy pulses, that are associated with the differentiated cell, that need to be countered) thereby initiating a countering process leading to its survival in the environment. As the environment fluctuates chaotically the DNA is responding continually to it and this is the life effect, and when a persistent change of environment slowly develops the DNA suffers controllable damage and undergoes a repair procedure resulting in extension to the DNA, new and adapted genes etc plus an extension to the differentiated type cells resulting in evolution of the species. (This process is such it should apply to members of a species who are subjected to very similar environmental conditions ensuring changing aspects are consistent and do not become conflicting in interbreeding)

To cope with the increasing mental environment the extensive associated radiant energy pulses have led to the "junk" DNA evolving in human cells ensuring survival. The survival aspects is the evolution of the protein that is injected into the synaptic connections of the glial brain cells enabling the incoming environmental energy pulses related to the senses to be transmitted, but in a muted state, allowing these cells to have an extended lifetime, hopefully for the life of the species. The pulsating response from the associated Junk nucleotide pairs of the DNA that in reality have been evolved with extensive RM capacity initiates the injection of protein specified by relevant genes into the synaptic gap that is then able to pass on the energy pulses that are synonymous with the demanding mental environment effect. The incoming effects from the various senses are funneled to a cell (neuronal) via the synapses where the total input is coordinated and memorized by the total pulsing energy effects causing recognizable distortions in the nucleotide pairs. This distortion effect, which is reliant on a physical property of materials that allows a deflection to recover at a given rate, provided it has not been strained beyond its elastic limit then accounts for the phenomenon of short term, medium term and long term memory that is recalled when the incoming radiant energy pulses reactivate the distorted nucleotides. (The extended "Junk" RM of the cell utilized in the glial cells of the brain) This is the process of conscious memory and depending on the numbers of times the memory effects have been installed or the intensity of the installation the memory effect may be that weak it is not recoverable or the memory may be intense and long lasting to permanent. (This accounts for PTSD to the degree that the distortions become permanent, resulting in a gene extension (evolution) and with an output production of a reactive chemical that affects the DNA of the reproductive cells resulting eventually in a similar gene change and this may result in the problem being inherited by any offspring). Survival output functions necessary are then activated,

THE SCIENCE OF LIFE AND EVOLUTION

such as responses of movement, speech etc. from the final receptive (neuronal) cell in an integrated feed system from the various senses. The protein, produced and injected into the synapses of the concerned cells to facilitate this flow of information has evolved as a response to coping with the radiant energy pulses related to normal environmental events requiring physical reactions and is not capable of dealing with any radiant energy pulses emanating from outside of this range.

The RM (Junk) has been evolved in a reaction to persistent incoming radiant energy associated with a requirement for reactions to particular behavioral events leading to individualistic sets of nucleotide pairs that are predisposed to similarly react to these incoming environmental events in the offspring i.e. inherited behavior patterns.

Conclusions

Without the process of protective protein production to control the period of survival of the DNA and hence the cell there would be no life.

The protective capability of a protein is an initiative of the evolved reaction by the DNA to a previously experienced persistent environmental energy pulse and as such it is produced to deal with it. In similar fashion aspects of the environment are dealt with and these energy pulses are involved in the production of specific proteins etc. however an aberrant radiant energy has not evolved a protection protein and the pulse passes through the existing protein, damaging the DNA.

Radiant energy associated with Smart Meters, Wind Turbines, Mobile Phones, Fluorescent Globes etc. have frequency and amplitude properties that are outside of the normal energy persistently experienced in the environment and as such, although regarded as "weak" have the potential to pass through the protective shield of the protein and cause damage to the DNA.

The potential damaging effect from theses sources is by primary means and therefore those cells that are directly exposed to the environment will experience the initial damage. Skin cells and brain cells will therefore be exposed to problems.

The Epigenetic effect is not and cannot be an activating entity. The DNA is the initiating entity of all protein etc due to the reaction of the Nucleotide pairs to the radiant energy of the environment, primary or secondary (the cause) initiating protein etc (the effect). All protein etc is produced to facilitate an effect (survival) and as such is utilized to serve various purposes. (A disastrous situation would occur with the two sources initiating activities and being at loggerheads, causing mayhem, in fact life would not have eventuated).

The evolutionary process results in each and every specimen of a species having similar but never the same DNA and hence as the protein produced is synonymous with the DNA it also is never the same. As one of the functions of task-orientated protein is to filter radiant energy pulses before accessing the DNA the various host specimens exposed to aberrant radiant energy pulses may demonstrate different reactions as this energy infiltrates to the DNA.

Members of a population may not be affected in the same manner when exposed to potentially the same energy pulses as conditions may vary e.g. live in a house lined with aluminium faced insulation.

N.B.

An example of the phenomenon of radiant energy pulses being obstructed or passed through a material due to their different frequency and amplitude is ultra violet rays are inhibited whilst infra red rays are readily passed through the molecules of a glass sheet.

PROTEIN, SMART METERS AND DOPE TAKING BY ATHLETES

◇◇

In the pre-life primeval swamps of the earth, as a range of physical elements became available chemical molecules formed, including sugar and the nucleotide acids, A, C, G and T. Eventually combinations of these molecules, exposed to naturally occurring enzymes came about forming complex molecules of RNA (half sides of DNA) with protrusions that were susceptible to bonding with Hydrogen atoms. The nucleotide acids C and G had protrusions susceptible to bonding with one and other, as did A and T via these H atoms. Eventually short lengths of the RNA molecules combined forming a restricted length molecule of DNA with some of the arrangements of the AC and GT nucleotide pairs harmonically responsive to the radiant heat energy pulses of the environment (The DNA molecule is receptive to vibration and resonating).

This vibration phenomenon occurred in a magnetic field provided by the presence of radioactive isotopes of carbon and potassium (C14 and K40). These elements are always present in the living cells of any species. As the nucleotide pairs resonated through the magnetic field an alternating current was generated in them and this caused the flexing nucleotide pairs to have a surrounding energy field that rose and fell in cohesion with the chaotic changes of the radiant energy pulses emanating from the environment. (After the individual RM's and genes were formed, with the additional environmental energy superimposed on the heat energy effect the phenomenon of Quantum Tunneling came into play, restricting the additional signaling effect to the relevant gene (s)) In the conditions the generated energy pulses from the DNA, with various chemical elements present, assisted a chemical reaction to form an enzyme that further resulted in assistance for a chemical reaction to form an outer enveloping layer of protein forming a cell membrane. Additionally the energy pulses emanating from the activated nucleotide pairs assisted enzymes to form, which accelerated a chemical reaction resulting in protein (chromatin) that coated the DNA molecule. This chemical coating did not necessarily shield the DNA from the damaging effects of the environment radiant energy with the arrangement of the amino acid molecules in the complex protein molecule and the frequency and amplitude of this radiant energy not clashing and therefore the radiant energy passed directly to the DNA, resulting in a repeat of this process, adjusting the nucleotide pairs (RM) that on resonance eventually guided the deliverance of a modified molecule of protein from the cell in a reaction to this incoming energy.

As the hit and miss repetition process eventually resulted in a protein molecule (chromatin) that did muffle the incoming radiant energy affording the DNA some protection from the damaging effects and as it did not exclude all of the energy effects from the nucleotide pairs of the DNA they eventually suffered fatigue at the hydrogen bonding joints and split apart, forming two molecules of RNA. An enzyme assisted chemical reaction then took place resulting in the formation of two additional RNA molecules that matched up with these highlighted existing RNA molecules and they bonded causing two duplicate molecule of DNA within individual cells to come into being. Also as a result of the whipping reaction of the molecule to the energy pulses the ends of the molecule met and joined. This established the circular DNA of the original single cell bacteria.

These chromatin proteins, then resulted in the extended life of the DNA and hence the cell until the DNA was eventually divided and chemically duplicated (Cell propagation) with the complete process being repeated for each individual cell.

In the course of time slight persistent variations to the environment occurred with consequential additions to the energy input. These variation penetrated the chromatin protein, and without the muffling effect of the amino acid molecules destructive damage occurred to several of the nucleotide pairs that were no longer harmonically responding resulting in an enzyme assisted chemical replacement of nucleotide pairs that were by the adversarial elimination process described above, once again in harmony with the energy pulses emanating from the environment causing the newly adjusted nucleotide pairs the Regulatory Memory, RM to resonate harmonically in response to this additional environmental energy. In conjunction with the developing gene(s) it resulted in the production of a modified output from the cell including further protective adjustment to the chromatin protein.

The gene came into being as during this process several of the adjoining nucleotide pairs were and are damaged, not destroyed but strained beyond their elastic limit and was no longer responsive, but represented the damaging effects of this aspect of the environment. For the DNA to survive in these damaging energy impacts a reactionary response was and is necessary and occurs.

The DNA molecule is made up of Nucleotide pairs, the numbers of which are extended as the DNA is exposed to a process of changing environmental energy damage and chemical reconstitution. When the resulting length of the DNA is extended the involved nucleotide pairs are susceptible to the continuing varying environmental radiant energy pulse that causes them to vibrate harmonically, switching on an initiating reaction from the associated gene that reflects that aspect of the environment resulting in the production of possibly protein, enzyme, hormone, vitamin, steroid etc from the cell that provide protection and survival techniques, within the environment, for the DNA of the cell and the associated specimen, hence survival of a species in its environment. (The concept of biologists that cells are self-propagating defies all logic and is a blot on their profession)

The protein etc produced in response to radiant energy impacts activating a gene is as individualistic to the specimen as the DNA, due to the initiating genes etc, having been evolved by the consistent but chaotic exposure of the individual's ancestors to persistent environmental changes. The changed exposure of a species where they are closely constricted to an environment is constant enough to develop characteristics that can eventually be inherited causing evolution of the species, making it overall compatible with its environment. The controlled reaction of the DNA to the potentially damaging energy pulses that are synonymous with the environment effect set up characteristics that allow the survival of the species. The process is ongoing over a specimen's lifetime.

This process is demonstrated by the incubation of a hen's egg where the DNA is first activated by the application of heat energy, activating the DNA where pluripotent cells are produced an array of embryo cells that all contain DNA with the appropriate genes etc, evolved over the species billions of years of history and passed on by the process of inheritance. During the propagation of these cells there is an uncommitted stage where a particular chemical randomly accesses them. These chemicals are found packaged within the egg and have been passed down having been individually evolved depending on the hen's ancestral evolutionary history. Normally the specimen's ancestral environmental history is similar but never quite the same. The chemicals themselves (Proteins, hormones etc) are produced, by differentiated cells that are a result of the specimens ancestors exposure to environmental effects (primary or secondary), evolving genes that when passed on by the inheritance process are an individual combination. This then accounts for the difference in individuals. The heat energy input activates the breakdown of these chemicals, that are randomly taken up by these uncommitted cells releasing controlled individual energy pulses of specific frequencies that activate the RM generating a current that causes a fluctuating energy field to activate genes etc that have been evolved as a result of the ancestors exposure to the responsible environmental energy pulse input. The initially pluripotent cells then become randomly reactive to these specific chemicals and then are differentiated cells. The result is then the specimen begins to evolve characteristics as the outputs of these cells are supported by the chemical makeup of the yolk resulting in further chemical outputs that then access the appropriate genes resulting in further differentiated cells that support the developing inherited characteristics of the chick. The process is ongoing with further application of energy pulses when the chick hatches resulting in the development of cells controlling characteristics that result in the survival of the chick and hence it's DNA in the environment.

Initially the production of the protein was a result of the damaging effect of the radiant energy pulse on the DNA causing a chemically activated reaction where nucleotide pairs were rebuilt and added to in an arrangement that were in harmonic tune with the radiant energy pulses thus lessening the tendency to damage them (The Regulatory Memory), whilst nucleotide pairs produced initially to the effects of an energy pulse that was the forerunner of the subject energy pulse were in a strained slack condition (The Gene) and as these nucleotide pairs having been initially evolved by the damaging effects of this energy they represented the environment by providing a template (memory) of the damage effect that was overcome leading to the DNA surviving. The dedicated RM representing the incoming effect of the environmental energy that is to be survived in is activated by this particular incoming energy effect as its chaotic application initiated the production of cells until their output provided sufficient response, when integrated with the other outputs from further differentiated cells for the survival of the DNA. The output of, proteins, enzymes, hormones, steroids, vitamins etc evolved supporting the DNA of the differentiated cells in a cohesive manner resulting in the cells outputs being compatible with the environment and therefore controlling the survival of the specimen.

With the advent of Eukaryotic species that are the result of one cell being invaded by another plus the addition of a mitochondria species that has the capacity to release stored environmental energy via secondary means (sugar) the cells became more efficient and complex arrangement of various internal cells evolved that utilized energy delivered by these indirect means i.e. the controlled reduction of the initiating sustenance chemical within the cells cytoplasm. This process resulted not only in the delivery of various chemical elements required for the production of the cells output but also the necessary energy for the initiation of the DNA reaction resulting in the cells production of these outputs. A further event was the cells becoming so overcrowded the DNA molecules broke up resulting in various lengths of DNA occurring and pairing up (The chromosomes pairs) in the cells nuclei. The whipping action that occurred as a result of the continuing energy impulses was the chromosome pairs were damaged at the ends fraying them. This further resulted in the formation of an enzyme (Telomerase) assisting in the development of a protein (Telomeres) forming a like bonding on the ends of the DNA thus preventing the destruction of the DNA chromosome by further fraying.

The process developed to release energy in controlled pulses within the cell with the secondary function of controlling the activation of the RM and therefore initiating a gene defined response from the cell e.g. a sugar molecule releases approx 140 controlled energy pulses when reduced in a cell and in the case of the evolved differentiated Pancreas cells this is directly implicated in the controlled production of Insulin, a complex protein ("Peptide"). The sustenance (sugar) is therefore acting as a conduit for the environmental energy effects responsible for its production. Various chemicals supported and provided by sustenance release a range of energy pulses of various frequencies on reduction and therefore control different outputs from the cells by way of initiating a reaction from the appropriate genes of a now differentiated cell.

With increasing evolutionary complexity various differentiated cells evolved with specific tasks of producing proteins, hormones, enzymes, vitamins, steroids etc dedicated to muscle fibre development, growth, energy outputs and in the case of mobile species mental outputs etc.

The attached document illustrates the helix trace (In the two dimensional view it appears to be a sine curve.) of the arrangement of Amino Acid molecules in a complex molecule of protein. It is also the trace of a radiant energy pulse and when associated with the pulse over and above the initial heat energy pulse from the developing environment that first accesses the DNA molecule the nucleotide pairs that were initially in harmonic tune to the radiant heat pulse suffers a damaging effect causing the straining of several of these nucleotide pairs and destruction of a further several nucleotide pairs a chemical reaction resulting in firstly the production of additional nucleotide pairs that once again was responsive to activating the production of a protein coating. In the event that the new nucleotide pairs activated a production of a protein that consisted of an arrangement of the amino acid molecules permitting the energy to pass through and further damage the DNA nucleotides the process was repeated. This process was and is repeated until a protein occurs (Scab like) that does muffle the effects of the incoming energy, bringing the activated output from the gene and hence the cell into near equilibrium with the environment effect (Evolution). (This resulted in the DNA molecules extended survival and by necessity this chemical process had the protective arrangement demonstrated in the attached document other wise DNA responses could not exist). The relevant gene etc of the DNA molecule continued to be accessed by the reduced effects of this energy pulse, being activated in response to the environment effects and producing protein etc. (Gradually evolving more complex responses with the passing of eons).

And so life began.

The range of proteins, hormones etc (Peptides) produced therefore is individually specific to the initiation of an output and having been evolved as a response to an environmental energy input providing survival by protection against damage, or a range of characteristic developments resulting in survival they are dependent on the specimen being exposed to particular energy effects to activate these outputs.

A variety of problems can exist with this process e.g. when the parents are from different environments sufficient evolutionary differences may be inherited resulting in diabetes 1. MS etc.

Autism is also involved in a slightly different fashion where the gut Clostridia bacteria being involved in the functioning of the brain in relation to sustenance is dysfunctionally evolved due to mans interference, leading to its eventual survival in an extreme change of diet or exposure to an antibiotic environment. The bacteria originally evolved being sustained by accessing the sugar produced by the forerunners of plant life (Single cell Lichen). The input of the sugar by the bacteria resulted in an output of Propionic acid. These molecules were recognized by the evolving eukaryotic species as indication sustenance was available as at this stage it had not yet evolved the senses or a brain but was reliant on a communication system between the cells. This system eventually evolved into the brain, neuronal capacity and senses.

A degree of reliance on the indication from the bacteria that suitable sustenance was available however remained as did a certain amount of control of movement, evolved to access the sustenance as the evolving species incorporated the bacteria into its makeup leading to the survival of this type of species (mammals).

The evolvement of the bacteria results in a dysfunctional output of propionic acid with changed isomeric molecules that on entering the brain cells release divergent energy effects that result in a confused output

Another matter of concern is, that as all chromatin proteins have been developed with a physical structure capable of muffling only a specific energy frequency and this is the energy that is dealt with by its particular differentiated cell. With the introduction of man induced energy frequencies that there is no shielding protection against, in the case of DNA protection, they can penetrate and bombard the DNA causing damage and mutations to it that are beyond its repair capacity, resulting in cancer etc. The introduction of smart meters, wind turbines, cell phones etc are all involved in this problem with their introduction of foreign radiant energy pulses.

An example of the phenomenon of radiant energy either penetrating or being deflected as described above is window glass passing infra red radiant heat, but rejecting ultra violet radiant energy, even although the ultra violet is regarded as the stronger of the two.

Finally does this situation mean that a diabetic person cannot play sport as they are taking a peptide (Insulin) injected into the stomach that results in improved performance? This situation can also impact on controlled diets and it can also be involved if say for instance, the person suffers from a weakness such as low muscle development and they take an unusual diet or a substitute prescribed muscle builder. Also where does legitimate medical treatment end and "Dope taking" take over? Further what is a Peptide drug?

Before further decisions are taken this whole process, which is the basis of life should be completely understood, before any more confused findings are bandied around.

www.ingramcontent.com/pod-product-compliance
Lightning Source LLC
Chambersburg PA
CBHW030941180526
45163CB00002B/663